Safa Rguez
Majdi Hamammi
Ibtissem Hamrouni Sallami

Advanced Plant Physiology: From Root to Flower

Safa Rguez
Majdi Hamammi
Ibtissem Hamrouni Sallami

Advanced Plant Physiology: From Root to Flower

A simplified approach

ScienciaScripts

Imprint

Any brand names and product names mentioned in this book are subject to trademark, brand or patent protection and are trademarks or registered trademarks of their respective holders. The use of brand names, product names, common names, trade names, product descriptions etc. even without a particular marking in this work is in no way to be construed to mean that such names may be regarded as unrestricted in respect of trademark and brand protection legislation and could thus be used by anyone.

Cover image: www.ingimage.com

This book is a translation from the original published under ISBN 978-620-6-70665-6.

Publisher:
Sciencia Scripts
is a trademark of
Dodo Books Indian Ocean Ltd. and OmniScriptum S.R.L publishing group

120 High Road, East Finchley, London, N2 9ED, United Kingdom
Str. Armeneasca 28/1, office 1, Chisinau MD-2012, Republic of Moldova, Europe
Printed at: see last page
ISBN: 978-620-7-65776-6

Advanced Plant Physiology: From Root to Flower

By
RGUEZ Safa
HAMMAMI Majdi
HAMROUNI SELLAMI Ibtissem

Table of contents

Foreword

In a world where time seems to be accelerating and every day brings its own new discoveries and challenges, reading remains a timeless refuge, an open door to infinite worlds of learning and reflection. It is in this spirit that I invite you to dive into the pages of this book, an oasis of knowledge and exploration, where curiosity is our compass and discovery our reward.

In our never-ending quest for understanding and meaning, books remain faithful companions, guiding us through the twists and turns of history, science, philosophy and the imaginary. They are beacons in the ocean of ignorance, illuminating our path with the light of knowledge and wisdom accumulated over the centuries.

This book you hold in your hands is much more than just a collection of words printed on paper. It's a journey, an intellectual adventure that will take you to distant lands and profound ideas, inviting you to explore the hidden treasures of the human mind and the unfathomable mysteries of the universe.

As you turn the pages, you'll discover captivating stories, insightful analyses and unique perspectives on a multitude of subjects. Whether you're a seeker thirsty for knowledge, a dreamer looking for inspiration or a traveler of the mind eager for adventure, there's something in this book to capture your imagination and expand your horizons.

So may this book become your companion on the road, a guide to new horizons and an inexhaustible source of wonder and reflection. May its pages inspire, educate and enrich you, and may they nourish your mind and soul long after you've closed its cover.

May this be the beginning of an unforgettable journey, an invitation to explore the wonders of the world and the depths of your own being. And may each page you turn bring you a little closer to the truth, beauty and grandeur that surround us all.

Bon voyage, dear reader, and may this literary adventure bring you as much joy and enrichment as it has brought me in creating it.

Kind regards,

Dr RGUEZ Safa

Introduction

In the complex interweaving of the plant kingdom, advanced plant physiology stands out like a golden key, unlocking the mysterious doors to the biological processes that drive plant life. Like an ecological poem, this scientific field reveals the subtle intricacies that guide the growth, reproduction and interaction of plants with their environment. Thus, "Advanced Plant Physiology: From Root to Blossom" is a captivating expedition into the very heart of this botanical symphony.

In this quest for knowledge, we are intrigued by the crucial importance of understanding physiological processes to cultivate in-depth understanding and, by extension, optimize plant growth. Far from being merely static entities, plants are dynamic beings, responding exquisitely to a complex ballet of internal and external signals. This demanding biological dance, orchestrated at cellular, tissue and organic levels, is both a work of art and a challenge to demystify.

The first step in this scientific odyssey is an in-depth exploration of the context of advanced plant physiology. This context is woven into the threads of photosynthesis, cellular respiration, transpiration, and a myriad of cellular processes that sustain plant life. Beyond the simple observation of external growth, it is the understanding of these intimate processes that enables us to grasp the true nature of plants and accompany them as they flourish.

The book is committed to unveiling this fascinating complexity by offering an in-depth look at the various aspects of plant physiology. From roots that burrow into the soil in search of nutrients to unfurling leaves that trap light for photosynthesis, each chapter details the intricacies and adaptations of the essential physiological processes that shape the course of plant life.

Beyond the pure exposition of facts, the book aspires to incite wider reflection on the importance of plant physiology in our world. The practical implications of this knowledge extend to the fields of agriculture, botany, pharmacology and beyond. How can we, as responsible stewards of the planet, integrate this knowledge to foster a more harmonious coexistence with the plant world and, by extension, with our own ecosystem?

So, embark with us on a journey where science and poetry meet in the furrows of roots, where leaves tell stories of photosynthesis, and where flowers are ephemeral poems in the great book of plant life. This introduction merely reveals the surface of a rich and captivating exploration that promises to enlighten, inspire and transform our understanding of advanced plant physiology. Welcome to this garden of learning, where each page is a leaf that unfolds to reveal the mysteries of plant life.

A. Fundamentals of Plant Physiology

1. Defining plant physiology and its role in plant development

Plant physiology is the branch of biology that focuses on the study of the vital processes and mechanical functions that govern plant development, growth and reproduction. It aims to understand how the different parts of a plant, from cells to organs, interact to sustain life and enable adaptation to various environments (Zavafer et al., 2023).

Plant physiology plays a fundamental role in plant development. It explores the molecular, cellular and tissue mechanisms that enable plants to perform vital functions, such as photosynthesis, respiration, nutrient uptake, water and substance transport, growth regulation, reproduction and response to environmental stimuli (Jain, 2018).

1.1. Photosynthesis

Plant physiology studies the process by which plants convert solar energy into chemical energy, stored in the form of glucose, through photosynthesis. This chemical reaction is crucial for the production of food and energy for the plant (Ferry & Ward, 1959).

1.2. Breathing

It also examines the process of respiration, where plants release the energy stored in glucose to fuel their metabolic processes. Plant respiration is essential for growth and the maintenance of cellular functions (Ferry & Ward, 1959).

1.3. Nutrient absorption

Plant physiology analyzes how plant roots absorb essential nutrients from the soil, such as nitrogen, phosphorus and potassium, necessary for growth and development (Ferry & Ward, 1959).

1.4. Transport of water and substances

It explores the mechanisms by which water, nutrients and other substances are transported through the plant, including transpiration and sap flow. (Ferry & Ward, 1959).

1.5. Growth and Development

Plant physiology examines the processes involved in regulating cell and tissue growth, as well as the factors influencing the development of plant organs and structures. (Ferry & Ward, 1959).

1.6. Responses to Stimuli

It studies how plants detect and respond to environmental stimuli such as light, gravity, temperature and chemical signals, in order to adapt to changing conditions (Ferry & Ward, 1959).

In short, plant physiology provides an essential analytical framework for understanding the complex biological mechanisms that enable plants to survive, grow and reproduce in a variety of ecosystems. Its role extends beyond the simple observation of plants, contributing to the development of sustainable agronomic strategies, plant biotechnology and the resolution of environmental and food security challenges.

2. Study of the basic structures of cells, tissues and organs

The study of basic structures in plant physiology is a fundamental pillar for understanding the complex functioning of plants at different levels of organization, from the individual cell to specialized organs. This in-depth exploration of basic structures provides insight into the harmonious coordination required for plant life, and offers a glimpse into the amazing adaptations that enable plants to cope with diverse environments (Brett & Waldron, 1996).

2.1. Plant Cells

Cell wall: Plant physiology focuses on the cell wall, a distinctive feature of plant cells, providing support and protection while influencing exchanges with the environment (Showalter, 1993).

Chloroplasts: Analysis of chloroplasts highlights their crucial role in photosynthesis, converting sunlight into chemical energy and producing oxygen (Showalter, 1993).

Vacuole: The large vacuole, central to many plant cells, is studied for its storage, waste degradation and osmotic regulation functions. (Showalter, 1993).

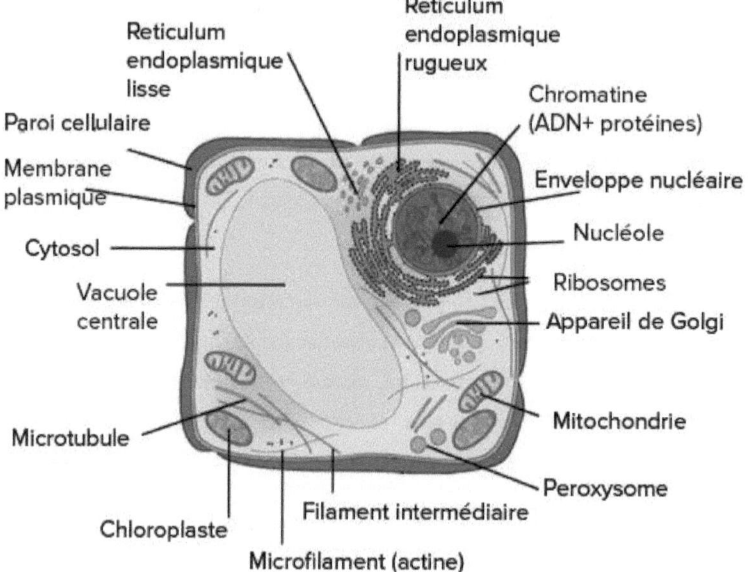

Figure 1. Diagram of a typical plant cell

(Review of plant vs. animal cells (lesson) | Khan Academy, n.d.)

2.2. Fabrics Plants

Meristems: The study of meristems, undifferentiated tissues, sheds light on the understanding of plant growth and development (Evert, 2006).

Conductive tissues Physiology explores the conductive tissues, xylem and phloem, which facilitate the transport of water, nutrients and photosynthetic products throughout the plant (Evert, 2006).

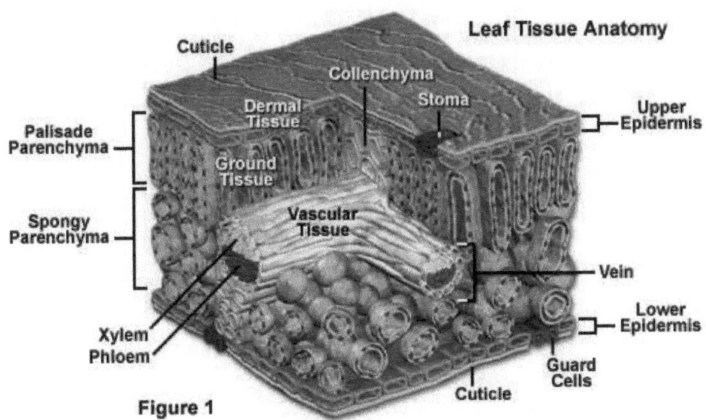

Figure 2. Plant tissue diagram
(*Molecular Expressions Cell Biology: Plant Cell Structure - Leaf Tissue Organization*, n. d.)

2.3. Plant organs

Roots Analysis of roots examines their adaptive structure for nutrient uptake and soil anchorage.

Stems Physiology explores stems as organs of support and transport, with a particular focus on nodes, internodes and buds.

Leaves The study of leaves reveals their specialized anatomy, involved in light capture, photosynthesis and transpiration regulation.

Each level of organization, from cell to organ, is intrinsically linked in the harmonious functioning of a plant. In-depth study of these basic structures enables us to grasp the fundamentals of plant physiology and to understand how plants respond to environmental stimuli, coordinate their growth and ensure their survival under a variety of conditions. This fundamental knowledge provides the basis for fields such as plant genetics, agronomy and biotechnology, paving the way for significant advances in the understanding and manipulation of vital plant processes.

3. Principles of photosynthesis and cellular respiration

The principles of photosynthesis and cellular respiration are at the heart of the energy processes that govern plant life. These two complementary mechanisms enable plants to generate energy from external sources, transforming sunlight into usable energy molecules and converting these molecules into the energy required for cellular functions. This energy symbiosis between photosynthesis and cellular respiration is essential to plant survival and growth (Rabinowitch, 1949).

3.1. Photosynthesis

Photosynthesis is a complex biological process by which plants convert light energy into chemical energy stored in organic molecules, mainly glucose. The main points to consider are :

Light Capture: Chlorophyll pigments in chloroplasts absorb sunlight, initiating the photosynthetic process (Yoshihara & Kumazaki, 2000).

Light reactions: within the thylakoids, light reactions separate water into oxygen, protons and electrons. The energy released is used to synthesize ATP and NADPH (Yoshihara & Kumazaki, 2000)..

Calvin cycle: In the chloroplast stroma, the Calvin cycle uses ATP and NADPH to fix carbon dioxide and produce organic molecules, mainly glucose (Yoshihara & Kumazaki, 2000).

Global Equation: The global equation for photosynthesis is $6\ CO_2 + 6\ H_2O +$ light $\rightarrow C_6H_{12}O_6 + 6\ O_2$ (Yoshihara & Kumazaki, 2000).

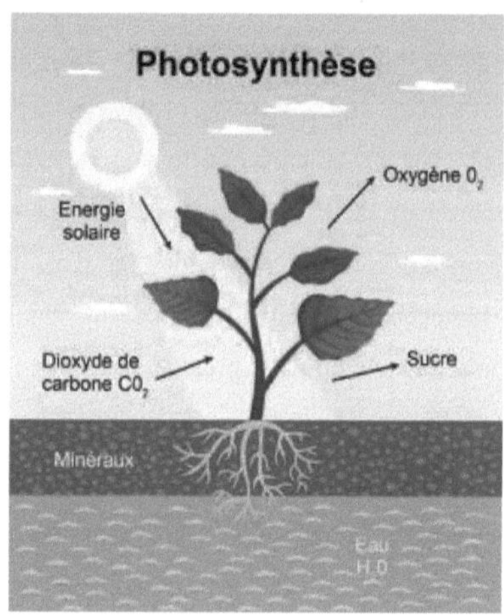

Figure 3. Photosynthesis
(*Photosynthesis*, 2020)

3.2. Cellular respiration

Cellular respiration is a process in which plants (and other organisms) break down organic molecules to release stored energy, usually in the form of ATP. Key steps include (Songer & Mintzes, 1994):

Glycolysis: In the cytoplasm, glycolysis breaks down glucose into two pyruvate molecules, producing a small amount of ATP.

Krebs cycle: In mitochondria, the Krebs cycle oxidizes pyruvate, releasing electrons which feed the electron transport chain.

Electron Transport Chain : Electrons moved along this chain generate energy used to pump protons across the mitochondrial membrane.

ATP synthesis: protons return to the mitochondrial matrix via ATP synthase, producing ATP.

Global equation: The global equation for cellular respiration is $C_6H_{12}O_6 + 6\ O_2 \rightarrow 6\ CO_2 + 6\ H_2O$ + energy (in the form of ATP).

These two processes are interconnected: photosynthesis produces the fuel needed for cellular respiration, while the latter provides the energy required for cellular processes, including photosynthesis. Thus, the symbiosis between photosynthesis and cellular respiration enables plants to maintain a dynamic energy balance to support growth and survival.

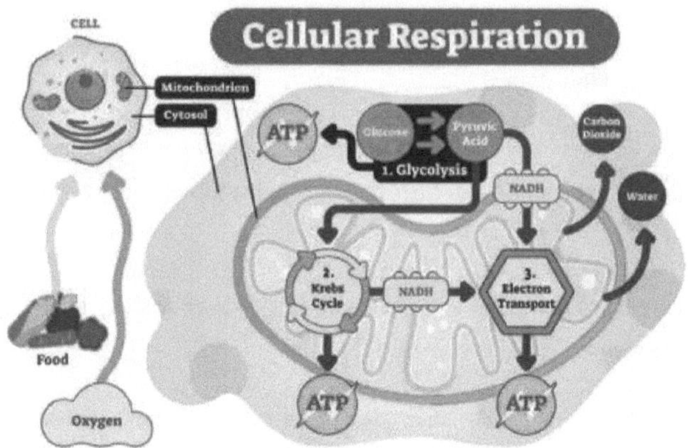

Figure 4. Cellular Respiration

(Cellular respiration - Definition and Examples - Biology Online Dictionary, 2023)

B. The Root and Nutrient Absorption

1. Root structure and function

Plant roots are essential structures that play a fundamental role in anchoring plants in the soil, absorbing nutrients and interacting with the environment. The complex structure of roots is adapted to various vital functions necessary for plant growth and survival (Barber & Silberbush, 1984).

1.1. Root structure

Cap (or *Calyptra*): At the root tip, the cap protects the active growth zone called the meristem, ensuring protection when penetrating the soil.

Growth Zone (or Meristem): This region, just behind the cap, is responsible for root growth in length. Meristematic cells divide actively, generating new tissue.

Radicular hairs: Radicular hairs are tiny, single-celled projections along the root surface. They increase the absorption surface, facilitating the uptake of water and nutrients.

Epidermis: The outer layer of the roots, the epidermis, is made up of protective cells that regulate nutrient absorption and gas exchange.

Cortex: The cortex is the layer of tissue between the epidermis and the central cylinder. It stores nutrient reserves and facilitates radial transport of substances.

Central cylinder (or Stele): At the center of the roots is the central cylinder, which contains the xylem (water transport) and phloem (nutrient transport).

Endoderm: The inner layer of the cortex, called the endoderm, acts as a selective barrier, controlling the passage of substances to the central cylinder.

Pericycle: This layer gives rise to lateral roots and can differentiate into additional vascular tissue.

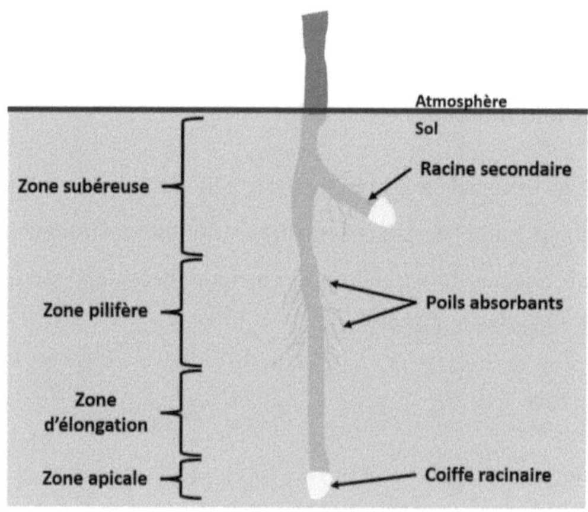

Figure 5. The different zones of a taproot
(*The root system*, s. d.)

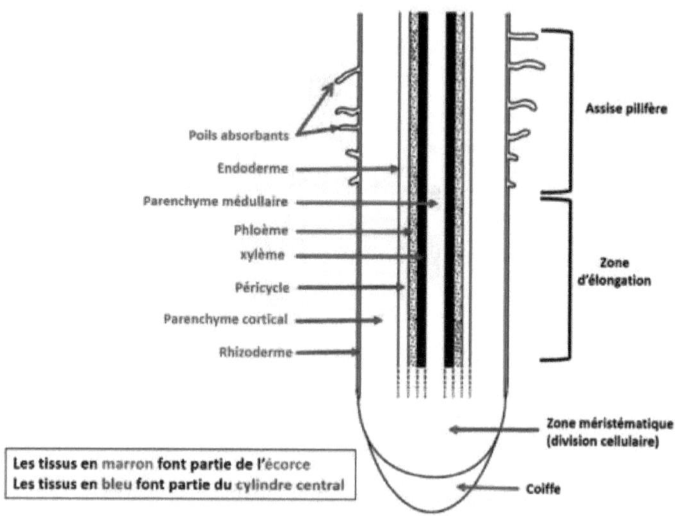

Figure 6. Internal primary root structure (monocot)
(*The root system*, s. d.)

1.2. Root functions

Nutrient absorption: Roots actively absorb water and essential soil nutrients, such as mineral ions, required for plant growth and development (Hodge et al., 2009).

Soil anchorage: Roots anchor the plant in the soil, providing stability and structural support, particularly important for growing plants (Hodge et al., 2009).

Nutrient storage: The root cortex can serve as a storage area for reserves of carbohydrates, lipids and other nutrient compounds, ensuring a constant supply of energy (Hodge et al., 2009).

Interaction with Microorganisms: Roots interact with soil microorganisms, establishing beneficial symbioses, such as mycorrhiza, that enhance nutrient uptake (Hodge et al., 2009).

Role in Environmental Signaling: Roots can sense environmental signals such as gravity, soil moisture and the presence of chemicals, thus regulating growth and development (Hodge et al., 2009).

Lateral root formation: The pericycle enables the formation of lateral roots, extending the absorption zone and improving the plant's overall efficiency (Hodge et al., 2009).

In short, root structure and function are closely linked, making a crucial contribution to the survival and prosperity of plants by enabling them to adapt to soil conditions and take advantage of available resources.

2. Soil nutrient uptake process

The process of soil nutrient uptake by plant roots is a key aspect of plant physiology. This absorption is crucial for the growth, development and maintenance of plant metabolic functions. The process of nutrient uptake involves several complex and coordinated steps at root level. Here's a detailed explanation of this process:

2.1. Root Substance Release

Roots release hydrogen ions (H^+) into the soil in a process called rhizodeposition. This acidifies the rhizosphere, the soil zone directly surrounding the roots (El Mekdad, 2023).

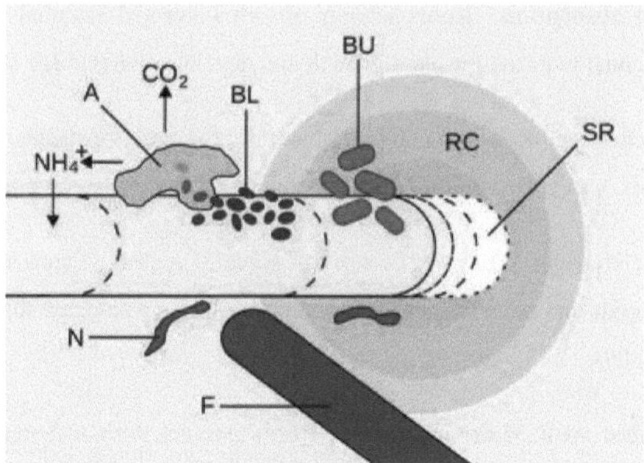

Figure 7. Rhizodeposition: secretion of carbon into the rhizosphere. ("Rhizodeposition", 2023)

2.2. Acidification of the Rhizosphere

Acidification of the rhizosphere facilitates the dissolution of soil minerals into ions, making them available for uptake by the roots (El Mekdad, 2023).

2.3. Nutrient Absorption by Radicular Hairs

Root hairs, tiny single-celled projections on the roots, increase the absorption surface. They are responsible for the active uptake of soil nutrients, such as mineral ions (nitrate, phosphate, potassium, etc.). (El Mekdad, 2023).

2.4. Active Transportation

The process of nutrient uptake is often an active transport, requiring energy to move ions against their concentration gradient. Ion pumps in the root cell membrane are involved in this root process (El Mekdad, 2023).

2.5. Selection and Specificity

Roots have a certain specificity when it comes to absorbing nutrients. Selective ion channels allow the selective entry of certain ions according to the needs of the root plant (El Mekdad, 2023).

2.6. Mycorrhizae

Symbiotic associations with mycorrhizal fungi can facilitate nutrient uptake. The fungal hyphae extend the zone of absorption and facilitate the exchange of nutrients between the plant and the root soil (El Mekdad, 2023).

2.7. Radial transport

Absorbed nutrients pass through the root cortex and reach the central cylinder, where they are transported upwards via the root xylem (El Mekdad, 2023).

2.8. Interaction with the endoderm

The endodermis, a layer of impermeable cells at the base of the cortex, regulates the passage of nutrients to the central cylinder, ensuring selective root uptake (El Mekdad, 2023).

2.9. Translocation in the plant

Once in the central cylinder, nutrients are transported through the xylem to the various parts of the plant, where they are used for growth, metabolism and other vital root processes (El Mekdad, 2023).

This complex process of nutrient uptake demonstrates how plants adapt to maximize the use of resources available in their environment. Precise coordination between the various root structures and functions is essential to ensure an adequate supply of nutrients for the plant.

3. Interaction with soil microorganisms

The interaction between plant roots and soil microorganisms is a fundamental aspect of plant physiology. These interactions are diverse and complex, playing a crucial role in plant nutrition, soil health and even disease resistance. Here are some of the main forms of interaction between roots and soil microorganisms (Kolb et al., 2017) :

3.1. Mycorrhizae

Mycorrhizae are symbiotic associations between plant roots and mycorrhizal fungi. These fungi form hyphae that extend the root absorption zone, facilitating the uptake of nutrients, particularly phosphates, and increasing resistance to environmental stresses (Fortin et al., 2008)

Figure 8. Mycorrhizae
(*Mycorhize*, s. d.)

3.2. Rhizobium and nitrogen fixation

Certain microorganisms, such as bacteria of the Rhizobium genus, establish a symbiotic relationship with the roots of leguminous plants. These bacteria fix atmospheric nitrogen, converting it into a form that can be used by plants, thereby improving the nitrogen nutrition of host plants (Galiana, 1990).

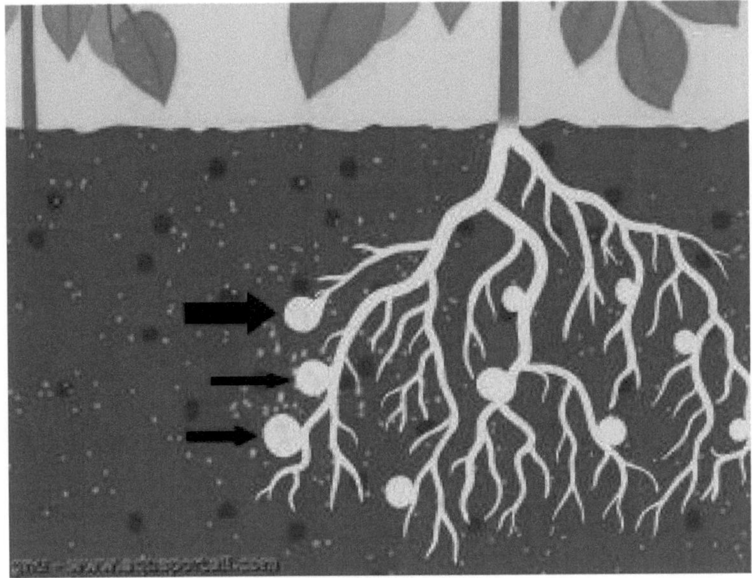
Figure 9. Nitrogen fixation
(*Nitrogen fixation*, n.d.)

3.3. Root exudates and signaling

Roots release root exudates, organic compounds such as sugars and acids, which serve as nutrients for soil microorganisms. In turn, these microorganisms can release compounds beneficial to the plant, and this interaction creates a specific rhizospheric environment (Mazziotti, 2017).

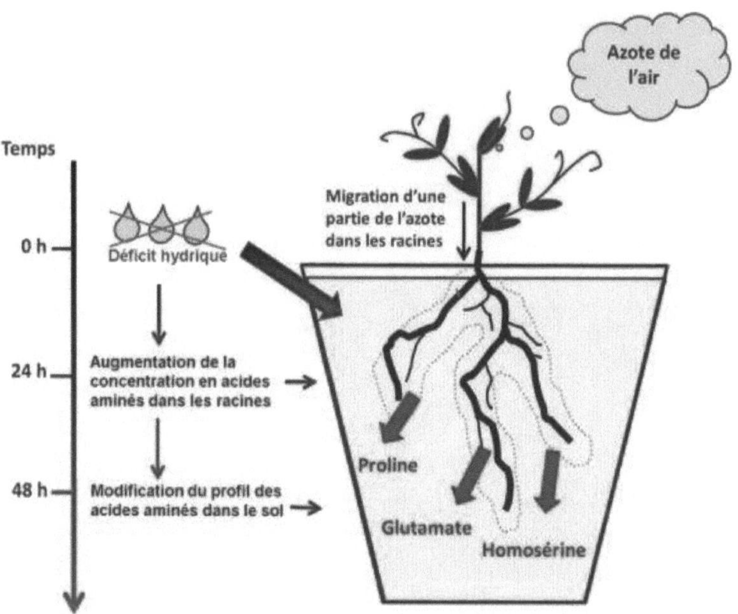

Figure 10. Changes in pea root footprint in response to water shortage (*Modification of pea root imprint in response to water shortage*, n.d.)

3.4. Pathogen Protection

Certain soil microorganisms, such as bacteria and fungi, can act as biological control agents, protecting roots against pathogens. They can colonize roots and produce antimicrobial compounds that limit pathogen growth (Wehner et al., 2010).

3.5. Organic matter decomposition

Microorganisms break down organic matter in the soil, releasing nutrients that become available to plant roots. This decomposition contributes to nutrient cycling in the soil (Wehner et al., 2010).

3.6. Root Growth Stimulation

Some soil microorganisms can stimulate root growth by producing plant growth hormones or improving nutrient availability (Belhadi & Zillal, 2020).

3.7. Chemical Signal Exchange

Complex chemical signals, such as volatile compounds, can be exchanged between roots and microorganisms, facilitating communication and modulation of adaptive responses.

3.8. Modulation of Stress Tolerance

Interactions with certain microorganisms can improve plant tolerance to abiotic stresses such as drought, high salinity or heavy metals.

These interactions create a dynamic microcosm called the rhizosphere, the region of the soil directly influenced by roots. Understanding these interactions is crucial to optimizing plant health, soil fertility and the sustainability of agricultural systems.

C. The Stem and Substance Transport

1. Role of the stem in transporting water, nutrients and hormones

The plant stem plays a crucial role in transporting water, nutrients and hormones throughout the plant, facilitating growth, development and response to environmental stimuli. Here's how the stem engages in these essential functions:

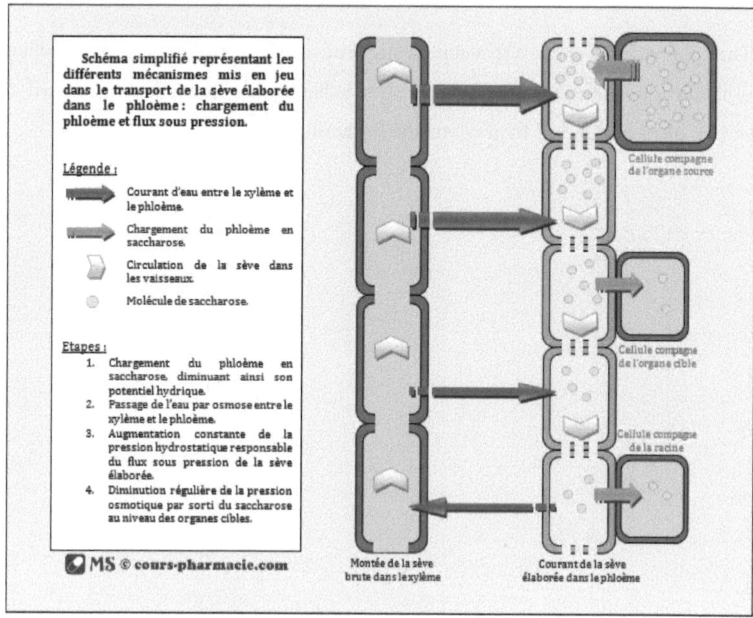

Figure 11. Water transport to xylem vessels
(SIMON, 2009)

1.1. Water transport (Xylem)

Xylem: The xylem is a vascular tissue specialized in transporting water and mineral salts from the roots to the aerial parts of the plant (Payette & Filion, 2018).

Ascent of Raw Sap: The stem allows the ascent of raw sap, which is essentially water with dissolved mineral salts, from the roots to the leaves(Payette & Filion, 2018).

Transpiration: transpiration, the evaporation of water by leaf stomata, creates a suction that draws soil water through the plant(Payette & Filion, 2018).

1.2. Nutrient transport (Phloem)

Phloem: The phloem is another specialized vascular tissue that transports organic nutrients, such as carbohydrates produced by photosynthesis, from leaves to other parts of the plant. (Hopkins, 2003).

Descent of Elaborated Sap: Elaborated sap, rich in organic nutrients, descends from the leaves to the roots and other parts of the plant. (Hopkins, 2003).

1.3. Hormones and Chemical Signals

Hormone transport: Stems facilitate the transport of plant hormones, such as auxin, gibberellin and cytokinin, which play an essential role in regulating growth, cell differentiation and response to environmental stimuli (Delaporte, 2009).

Chemical communication: Stems enable chemical communication throughout the plant by carrying molecular signals, enabling efficient coordination of the various plant parts (Delaporte, 2009).

1.4. Structural support

Mechanical support: The stem provides mechanical support to the plant by maintaining a vertical structure, allowing optimal leaf arrangement for light capture and resisting external forces such as wind.

1.5. Nutrient Storage

Storage parenchyma: Certain stem tissues, such as cortical parenchyma, can store nutrients and water, acting as a reserve for the plant in times of need (Plavcová & Jansen, 2015).

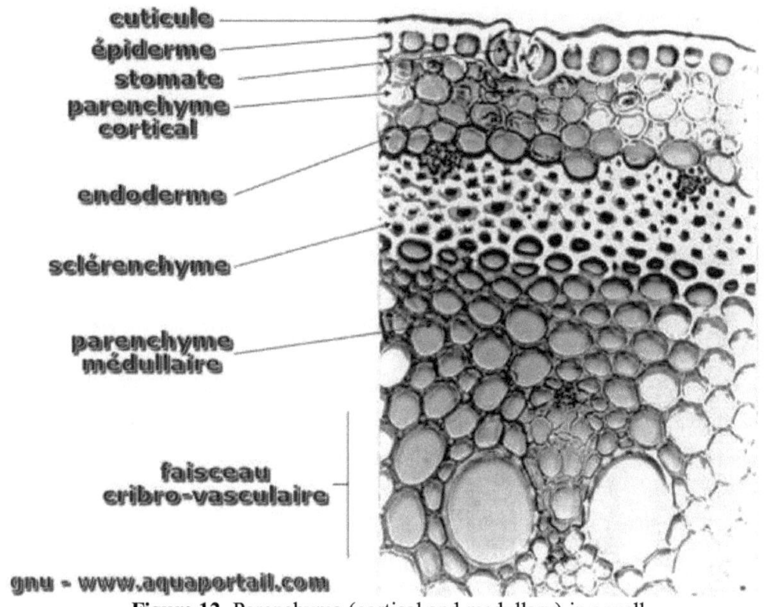

Figure 12. Parenchyma (cortical and medullary) in a wall
(*Parenchyma*, n.d.)

1.6. Phototropism

Response to Light: The stem can exhibit directional growth in response to light, a phenomenon known as phototropism, which enables the plant to optimize its exposure to light for photosynthesis. (Thimann, 1967).

In short, the stem acts as a multifunctional conduit for the efficient transport of vital substances throughout the plant. Its role in the vascular system, hormone regulation and the coordination of responses to environmental signals is essential for plant survival and prosperity.

2. Mechanisms of transpiration and sap pressure

The mechanisms of transpiration and sap pressure are two intimately linked processes that play a crucial role in the transport of water and nutrients through plants. Here's a detailed explanation of these mechanisms (Cruiziat et al., 2001) :

2.1. Sweating

Transpiration is the process by which plants lose water in the form of vapor through the stomata found mainly on leaves. Here are the main transpiration mechanisms:

Stomata opening and closing: Stomata are small pores on the leaf surface that control gas exchange and transpiration. They open and close to regulate water loss and carbon dioxide intake.

Water potential gradient: Water is evaporated from leaf cells into the atmosphere, creating a water potential gradient that draws water from adjacent tissues, then from the xylem vessels.

Water tension: Transpiration creates tension in the xylem vessels, called water tension, which pulls water from the roots to the leaves. This is due to the cohesion of water molecules and the adhesion of water to vessel walls.

Foliar transpiration: Most of the water lost through transpiration comes from the leaves, where the surface area is greater and stomata are more abundant. This process is essential to maintain the flow of water through the plant.

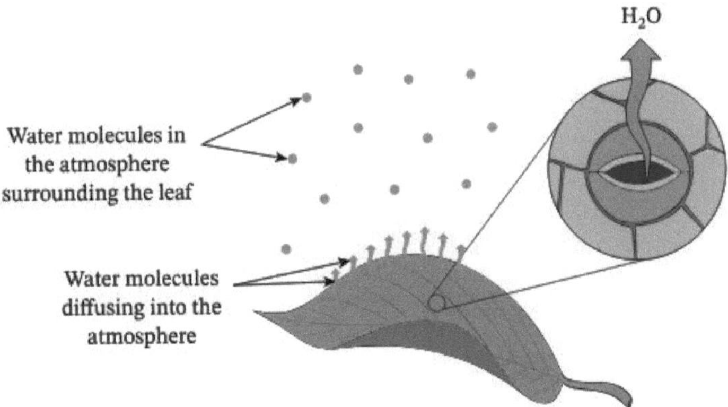

Figure 13. Diffusion of water vapour from stomata to the atmosphere through transpiration.

(Nagwa, n. d.)

2.2. Sap pressure

Sap pressure, also known as root pressure or turgor pressure, is a hydrostatic force generated by the absorption of water by roots and the ascent of sap in xylem vessels. The main mechanisms of sap pressure are as follows

Water absorption by roots: Roots actively absorb water from the soil by osmosis, thanks to the difference in solute concentration between root cells and the surrounding soil.

Cell turgor: Water absorption creates hydrostatic pressure, or turgor, in root cells, making them rigid and swollen.

Root Thrust: The turgidity of root cells exerts a force that pushes water upwards in the xylem vessels, against gravity, towards the aerial parts of the plant.

Positive sap pressure: Under certain conditions, this root thrust can lead to positive sap pressure, where water is forced out of leaves or stems, visible as guttation or sap oozing from wounds.

These mechanisms of transpiration and sap pressure ensure the efficient transport of water and nutrients throughout the plant, providing the resources needed for growth and metabolic function.

3. Structural adaptations to withstand environmental constraints

The plant stem plays an essential role in many structural adaptations that enable plants to withstand environmental stresses. Here are some of the ways in which stem structure contributes to these adaptations (Collin, 2001):

3.1. Resistance to strong winds

Flexible stems allow plants to bend under wind pressure without breaking. This flexibility is due to the stem's internal structure, which can contain softer, more elastic tissues.

3.2. Efficient water transport

The xylem vessels in the stem play a crucial role in transporting water from the roots to the aerial parts of the plant. Structural adaptations such as the arrangement of vessels and the presence of supporting tissues strengthen the stem, ensuring efficient water transport even under water-stressedconditions (Cruiziat et al., 2001).

3.3. Water storage

Some plants store water in their stems, particularly in arid environments. The succulent tissues of the stem can swell to store large quantities of water, enabling the plant to survive long periods without rain.

3.4. Light adaptations

Stem structure can also be adapted to maximize light absorption in shady environments. For example, some plants have slender, tall stems that allow them to rise above the canopy to reach the sunlight.

3.5. Resistance to herbivores

Some plants have structural adaptations in their stems to deter herbivores. For example, the presence of thorny or prickly structures on the stem can make the plant less attractive to foraging animals.

3.6. Fire response

In ecosystems subject to frequent fires, some plants have structural adaptations that enable them to survive fires. For example, thick underground stems or protected buds can survive high temperatures and grow back quickly after a fire.

In short, the stem structure of plants is closely linked to their adaptations to withstand environmental stresses. These structural adaptations enable plants to survive and thrive in a wide variety of conditions, from arid deserts to tropical rainforests.

D. Leaves and Photosynthesis

1. Leaf anatomy

Leaves, often considered the centers of plant life, are complex structures that host a multitude of vital processes, in particular photosynthesis. Understanding their anatomy is essential to fully grasp their function and role in plant life (van Lenteren & Ponti, 1991).

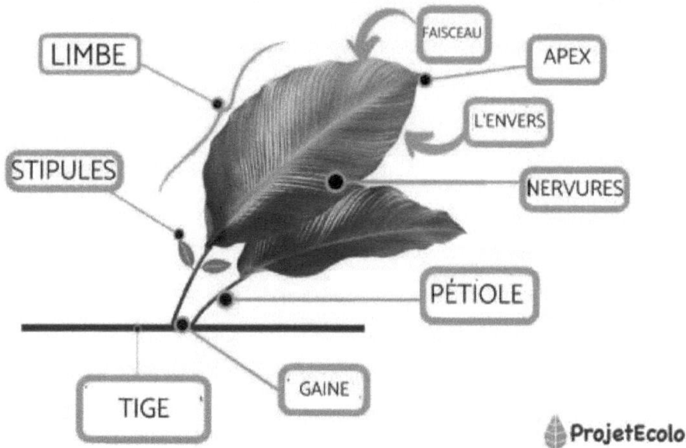

Figure 14. Anatomy of a leaf
(*Anatomy of a LEAF + explanatory DIAGRAM, n.d.*)

1.1. Epidermis

The epidermis, the outer layer of leaves, is generally made up of thin, transparent cells that act as a protective barrier against mechanical damage and excessive water loss. To reinforce this function, the epidermis is often covered with a waxy cuticle, a water-repellent layer that reduces water loss through evaporation while preventing the penetration of pathogenicmicroorganisms (van Lenteren & Ponti, 1991).

1.2. Stomata

Stomata, found in greater numbers on the underside of leaves, are microscopic structures made up of two guard cells surrounding an opening. These pores allow the exchange of gases essential for photosynthesis: carbon dioxide enters the leaf, while oxygen and water vapor leave. Regulating the opening and closing of stomata is essential for regulating the amount of water lost through transpiration and the amount of carbon dioxide available for photosynthesis(van Lenteren & Ponti, 1991).

1.3. Mesophyll

The mesophyll, the central zone of leaves, is where most photosynthesis takes place. It is divided into two distinct layers: the palisading parenchyma, located in the upper part, is characterized by tightly packed cells rich in chloroplasts, while the lacunar parenchyma, located in the lower part, is made up of looser cells that promote air circulation and gas diffusion (van Lenteren & Ponti, 1991)..

1.4. Conductive Vessels

The conducting vessels, xylem and phloem, transport the substances essential for photosynthesis to and from the leaves. The xylem transports water and minerals from the roots to the leaves, while the phloem transports the carbohydrates produced by photosynthesis to other parts of the plant for growth and storage (van Lenteren & Ponti, 1991).

1.5. Adaptations

Leaves display an incredible diversity of anatomical adaptations depending on the environmental conditions in which the plant grows. For example, in dry environments, leaves may be thick and fleshy to store water, while in shady environments they may be broad and thin to efficiently capture sunlight (van Lenteren & Ponti, 1991).

2. Photosynthesis processes and regulation of energy production

Photosynthesis is a vital process for plants, algae and certain bacteria, providing the energy required for growth, development and reproduction. Let's take a closer look at the various stages of photosynthesis and the regulatory mechanisms that ensure its efficiency (Smith et al., 1997):

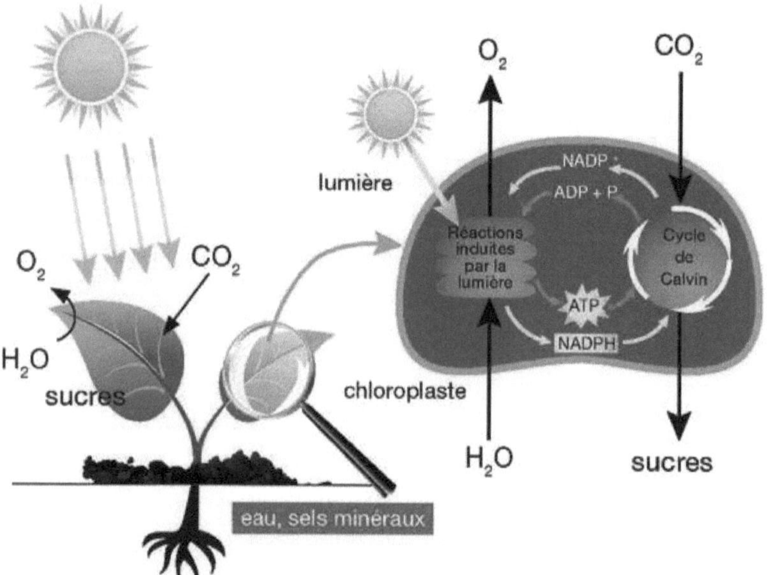

Figure 15. Diagram of photosynthesis at the leaf level
(Boussac & Mathis, 2015)

2.1. Light Capture

Chlorophyll pigments, mainly chlorophyll a and chlorophyll b, are responsible for capturing light. These pigments are found in the thylakoids, the inner membranes of chloroplasts. Light is absorbed by chlorophyll pigments and transferred to other photosensitive molecules in the process of light energy capture (Smith et al., 1997) .

2.2. Light reactions

Light reactions take place in the thylakoids and consist of two phases: photoexcitation and photodegradation of water. During photoexcitation, light energy absorbed by chlorophyll pigments excites electrons, which are transferred along an electron transport chain located in the thylakoid membranes. This electron transfer generates ATP and NADPH, which are used in the next phase of the process (J. Z. Zhang & Reisner, 2020).

2.3. Calvin Cycle (Obscure Reactions)

The Calvin cycle, the dark phase of photosynthesis, takes place in the stroma of chloroplasts. In this cycle, atmospheric CO_2 is fixed and converted into organic molecules, mainly glucose. CO_2 is first attached to a molecule of ribulose-1,5-bisphosphate (RuBP) by the enzyme rubisco, forming two molecules of 3-phosphoglycerate (3-PGA). These molecules are then converted into glyceraldehyde-3-phosphate (G3P) molecules, which are used to synthesize glucose and other organic compounds (Gurrieri et al., 2021)..

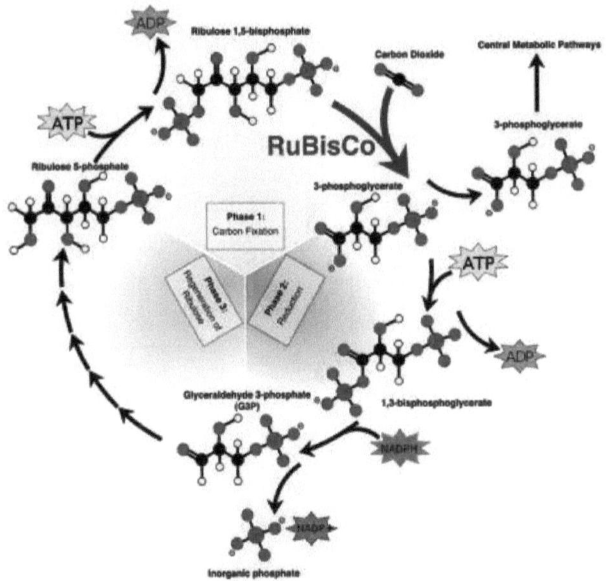

Figure 16. Calvin cycle (Obscure reactions)
(Futura, s. d.)

2.4. Power Generation Regulation

Photosynthesis is regulated by several factors to adapt to changing environmental conditions. For example, CO_2 concentration, temperature and water availability all influence the rate of photosynthesis. High levels of light can increase photosynthetic activity up to a point, but too much light can damage plant tissue through photoinhibition. Plants also have mechanisms to protect oxygen-sensitive enzymes, such as rubisco, from oxidative damage (Sirvydas et al., 2010)..

2.5. Energy use

The energy produced by photosynthesis is used for many of the plant's metabolic functions, such as the synthesis of carbohydrates, lipids, proteins and other organic compounds essential for growth and development. Some of this energy is also stored as starch in the chloroplasts for later use (Sirvydas et al., 2010)..

3. Leaf adaptations to maximize photosynthetic efficiency

Plant leaves have evolved to maximize their photosynthetic efficiency, enabling them to capture and make the best use of light energy to produce carbohydrates and other organic compounds required for growth and development (Zhu et al., 2010). Here are some of the main adaptations that leaves make to optimize their photosynthetic efficiency:

3.1. Shape and layout

Leaf shape varies according to species and habitat. Broad, flat leaves have a greater surface area exposed to light, increasing the plant's ability to capture light energy. Leaves arranged to minimize self-shading allow better light absorption by the lower leaves, maximizing the use of light energy in the canopy (Givnish, 1979).

3.2. Internal structure

The internal structure of leaves is adapted to optimize light absorption and gas exchange. Chloroplasts, which contain photosynthetic pigments, are concentrated in the mesophyll cells, making them accessible to light. In addition, the membranes of the thylakoids, where the light reactions of photosynthesis take place, are organized to increase the surface area for light capture (Oguchi et al., 2003)..

3.3. Photosynthetic pigments

Photosynthetic pigments, such as chlorophyll a and b, efficiently absorb light in different parts of the light spectrum. Carotenoids act as accessory pigments, capturing light in wavelengths where chlorophyll is less efficient. This combination of pigments enables plants to capture light efficiently under different lighting conditions (Y. Zhang et al., 2022).

3.4. Stomata

Stomata are key structures for the gas exchange required for photosynthesis. Stomata density and distribution vary according to species and environmental conditions. In dry environments, plants can regulate stomatal density to reduce transpiration while maintaining a sufficient supply of carbon dioxide for photosynthesis (Y. Zhang et al., 2022).

3.5. Adapted to water and high temperatures

The leaves of plants adapted to arid environments often have features that reduce water loss through transpiration. These may include sunken stomata, waxy epidermis or specialized structures such as trichomes. In addition, plants in hot climates may have adaptations to limit heat absorption, such as reflective leaves or hairs that reduce the amount of light absorbed (Berry & Bjorkman, 1980).

3.6. Photoprotection

Plants have developed mechanisms to protect photosynthetic pigments from damage caused by excessive exposure to light, a phenomenon known as photoinhibition. Carotenoids act as antioxidants, neutralizing the free radicals produced during photosynthesis. In addition, some plants can regulate the amount of light absorbed by changing the orientation of their leaves or by producing additional photoprotective pigments (H. Zhang et al., 2016).

These leaf adaptations demonstrate the incredible diversity and complexity of mechanisms that enable plants to optimize their photosynthetic efficiency in a variety of environments. By combining these adaptations with other metabolic processes, plants are able to adapt and thrive under changing environmental conditions.

Figure 17. Color changes in the leaves of *Physocarpus amurensis* Maxim and *Physocarpus opulifolius* "Diabolo" under natural and low light intensities. (H. Zhang et al., 2016)

E. Flowers and Reproduction

1. Flower structure and reproductive organs

1.1. Flower morphology

1.1.1. Perianth

The sepals, generally dark green, form the first protective envelope of the flower, known as the calyx. They may be simple or fused to form a tube or bell around the other floral organs. Sepals play a crucial role in protecting the reproductive organs during the flower bud phase, and may remain present after flowering. (Wilson & Just, 1939).

The petals, often colorful and fragrant, make up the corolla. Their main role is to attract pollinators by offering a visual and olfactory reward. Petal shapes, textures and colors vary enormously between species and can be adapted to attract specific pollinators, such as insects, birds or mammals (Wilson & Just, 1939).

1.1.2. Organ positioning

The positioning of the reproductive organs can have a significant impact on the plant's mode of reproduction. In some flowers, the stamens and pistil are positioned to encourage self-pollination, where pollen from one flower fertilizes the ovules of the same flower. In other cases, the reproductive organs are positioned to encourage cross-pollination, where pollen is carried from one flower to another by external agents, such as insects or wind. (Wilson & Just, 1939).

1.1.3. Inflorescence

Inflorescence refers to the arrangement of flowers on the plant. Flowers can be grouped in different configurations, such as racemes, clusters, umbels, capitula, etc. These arrangements can have an impact on the visibility of flowers to pollinators, on the ease of access to resources such as nectar and pollen, and on the way flowers are pollinated and dispersed (Wilson & Just, 1939).

1.1.4. Adaptations

Flowers display a multitude of adaptations to attract pollinators and ensure reproductive success. These adaptations can include nectar production to attract insects and other pollinating animals, specific color patterns to guide pollinators to the reproductive parts of the flower, attractive fragrances, specialized structures to provide resting or supporting surfaces, and mechanisms to prevent self-pollination or self-fertilization.

Flower morphology is not only a manifestation of nature's aesthetic diversity, it also reveals the complex and specialized adaptations flowering plants make to ensure their reproduction and survival in varied environments.

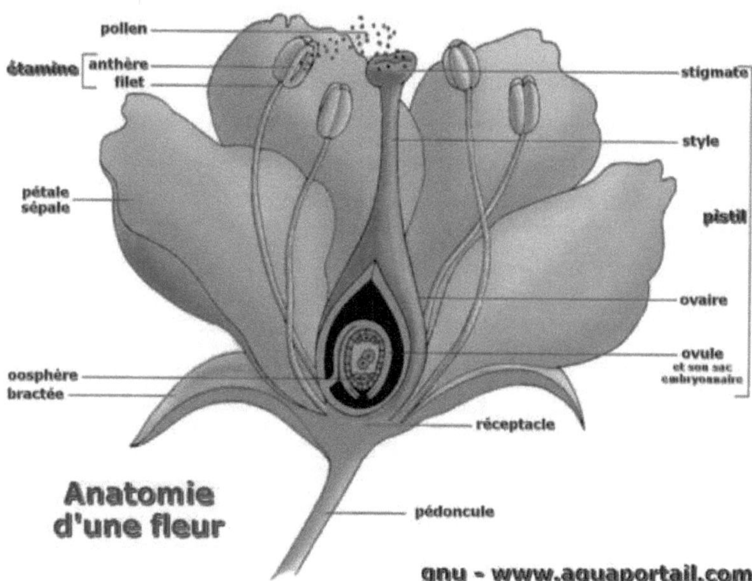

Figure 18. Schematic diagram of the anatomy of a flower.
(*Flower*, s. d.)

1.2. Reproductive organs

The reproductive organs of flowers are highly specialized structures that play an essential role in the sexual reproduction process of flowering plants (angiosperms). These organs enable the formation of seeds, thus ensuring the survival and propagation of plant species. Here's an in-depth exploration of the reproductive organs of flowers (Hermann & Kuhlemeier, 2011) :

1.2.1. Pistil

The pistil, also known as the carpel, is the female organ of the flower. It is generally located in the center of the flower and consists of several parts:

The stigma: This is the upper, often sticky, part of the pistil, specially designed to receive pollen. The stigma is covered with a viscous substance called liquid stigma, which helps the pollen adhere.

Style: An elongated structure connecting the stigma to the ovary. The style provides a passage for the pollen tube, allowing pollen to travel from the stigma to the ovary.

The ovary: This is the lower part of the pistil, containing the ovules. The ovary also contains the ovaries, which develop into seeds after fertilization. In some cases, the ovary transforms into a fruit after fertilization, protecting the developing seeds.

1.2.2. Stamens :

The stamens are the male organs of the flower, responsible for producing and releasing pollen. Each stamen is composed of two main parts:

The anther: This is the upper part of the stamen, where the pollen is found. The anther contains pollen sacs, which produce and release pollen grains. When the pollen is ripe, the anther opens to allow dispersal.

Filament: This is the elongated part of the stamen that connects the anther to the flower. The filament provides structural support for the anther and facilitates pollen release.

1.2.3. How it works

When flowers are ready to be pollinated, pollen is transported from the stamens to the stigma of the pistil. Once the pollen is deposited on the stigma, it germinates and develops a pollen tube through the style, eventually reaching the ovary. In the ovary, the pollen tube releases the male gametes, which fuse with the ovules, triggering the fertilization process. After fertilization, the ovules develop into seeds inside the ovary, and in many cases, the ovary transforms into a fruit to protect and disperse the seeds (Hermann & Kuhlemeier, 2011).

The reproductive organs of flowers demonstrate remarkable adaptation and specialization to facilitate the reproduction of flowering plants. Thanks to these complex, interdependent structures, plants can reproduce successfully and ensure the long-term survival of their species in a wide variety of environments and conditions (Hermann & Kuhlemeier, 2011)

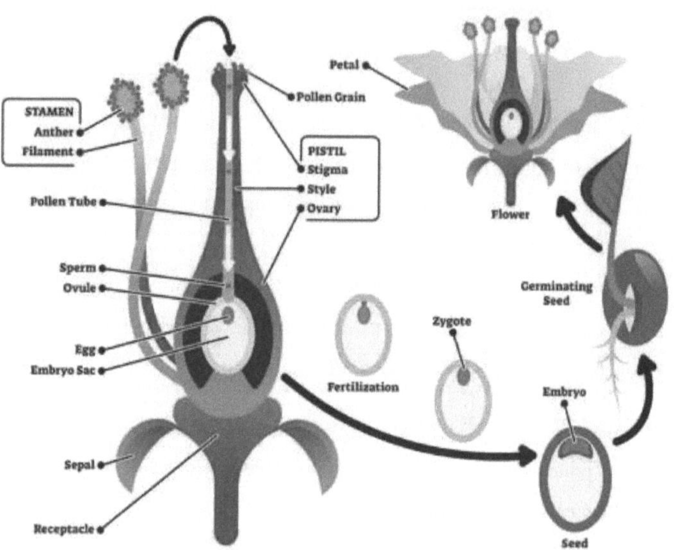

Figure 19. Sexual reproduction in flowering plants.
(Martin, 2022)

2. Pollination and Fertilization Mechanisms
2.1. Pollination

Pollination is a fundamental process in the reproduction of flowering plants, and can be achieved in a variety of ways, depending on the characteristics of the flower and its environment. Among the most common methods are wind pollination and insect pollination. Each of these methods has its own specific adaptations that favor successful pollination for the plants concerned (O'Neill, 1997).

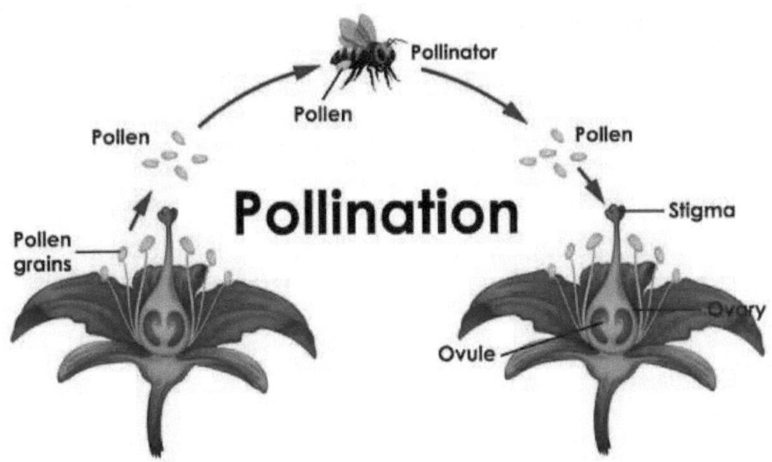

Figure 20. Pollination By The Flower And The Bee

2.1.1. Wind Pollination

Wind pollination, also known as anemophilia, is characteristic of flowering plants whose flowers are small, not very colorful and not very fragrant. These plants generally produce large quantities of light, dry pollen, which is easily carried by the wind over long distances. Flowers that depend on wind pollination often have features such as single flowers, exposed stamens and feathery stigmas to capture pollen. Common examples of wind-pollinated plants include grasses, cypresses and some flowering trees (Friedman & Barrett, 2009).

2.1.2. Insect pollination

Insect pollination, or entomophily, is a common pollination strategy for many flowering plants, especially those that attract insects with bright colors, scents and nectar. These plants often have large, colorful and fragrant flowers, designed to attract insect pollinators such as bees, butterflies, beetles and flies. Flowers can have a variety of attractive features, such as wide petals to provide a landing platform, nectaries to produce nectar, and patterns or visual guides to help insects find the reproductive parts of the flower. This beneficial interaction between flowering plants and insect pollinators is often mutually beneficial, providing food for insects while ensuring plant reproduction (Kevan & Baker, 1983).

2.1.3. Importance of Pollination

Pollination plays an essential role in the reproduction of flowering plants, contributing to seed and fruit production. It is also crucial for maintaining biodiversity and the health of ecosystems, as well as ensuring the pollination of agricultural crops and food production. The diversity of pollination methods reflects the adaptation of plants to a variety of environments and conditions, and underlines the importance of interactions between plants and pollinators for the survival and prosperity of life on Earth (Ratto et al., 2018).

2.2. Fertilization

Fertilization is a crucial stage in the reproductive process of flowering plants, taking place after pollination (Dumas, 2001). It involves the fusion of male and female gametes to form a new embryonic organism. Here's an in-depth exploration of the fertilization process:

2.2.1. Pollen germination

After being deposited on the stigma, pollen germinates to form a pollen tube. This pollen tube develops through the style, which connects the stigma to the ovary, guided by chemical signals and cellular growth reactions (sonika, 2023).

2.2.2. Moving gametes

Inside the pollen tube, the male gametes move towards the ovules located in the ovary. The male gametes are transported by the pollen tube to their final destination in the ovule (sonika, 2023).

2.2.3. Gamete Fusion

Once the male gametes reach the ovules, gamete fusion takes place. One of the male gametes fuses with the central nucleus of the ovule to form a diploid zygote, which is the beginning of the new plant embryo. Meanwhile, the other male gamete can fuse with other nuclei in the ovule to form nutrient tissue, which will provide the resources needed for the embryo to grow (sonika, 2023).

2.2.4. Embryo formation

The diploid zygote formed by the fusion of male and female gametes develops into an embryo inside the ovule. The embryo is the initial stage of the new plant, containing the genetic material inherited from both parents. It undergoes a series of cell divisions and developments to form the basic structures of the plant as it grows (sonika, 2023).

2.2.5. Nutrient tissue formation

In parallel with embryo development, the central nucleus of the ovum can fuse with other male gametes to form nutrient tissue. This nutrient tissue will provide the necessary nutrients for the growing embryo, enabling it to grow and develop in the future (sonika, 2023).

2.2.6. Importance of Fertilization

Fertilization is an essential stage in the life cycle of flowering plants, leading to the production of seeds and fruit. Seeds contain the embryo of the new plant, as well as the nutrient reserves required for its initial growth. Fruits protect the developing seeds and facilitate their dispersal, ensuring the propagation of plant species in their environment. Without the fertilization process, the reproduction of flowering plants would be impossible, compromising biodiversity and the survival of many plant species (sonika, 2023).

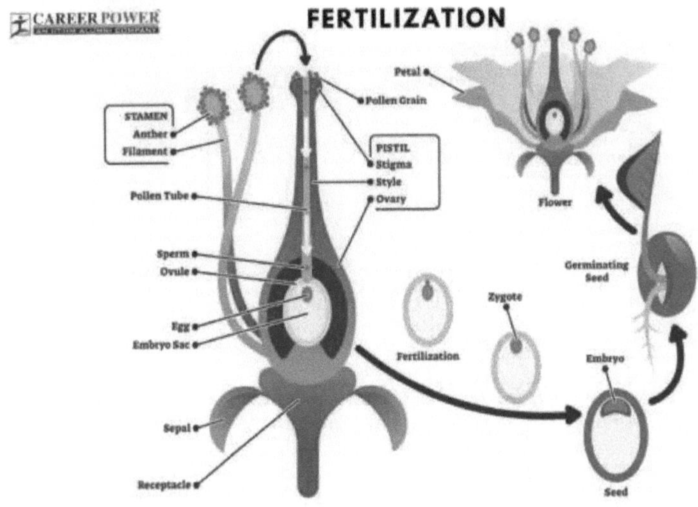

3. Seed and fruit development

3.1. Seed formation :

Seed formation is a crucial stage in the reproductive process of flowering plants (angiosperms), occurring after fertilization (Guo et al., 2022). This stage marks the beginning of the development of the next generation of plants. Here's a detailed exploration of the seed formation process:

3.1.1. Development of the Ovary into Fruit

After fertilization, the ovary of the flower develops into the fruit. The fruit can come in a wide variety of shapes, sizes and textures, depending on the plant species concerned. Its main function is to protect the developing seeds and facilitate their dispersal in the environment (Dennis Jr., 1984).

3.1.2. Transformation of the egg into a seed

Inside the fertilized ovary, the ovule transforms into a seed. The ovule contains the embryo of the new plant, as well as nutrient storage tissues and a protective shell (Dennis Jr., 1984).

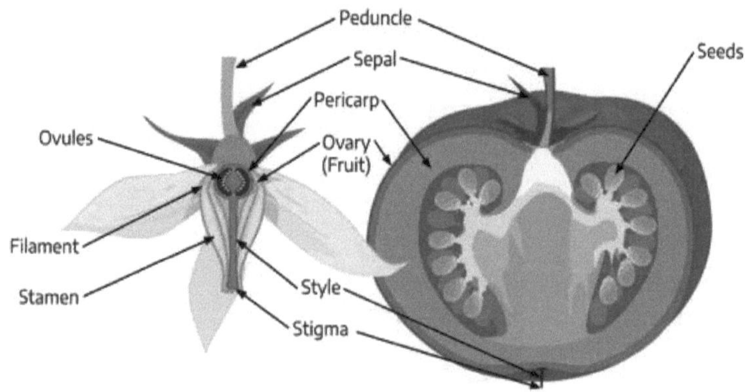

Figure 22. Tomato flower and fruit.
(Michaels et al., 2022)

3.1.3. Seed composition

The seed is made up of several essential elements:

Embryo: This is the young, developing plant resulting from the fusion of male and female gametes. The embryo contains the basic structures necessary for future plant growth, such as the radicle (future root), hypocotyl (future stem) and cotyledons (embryonic leaves). (Barthet & Daun, 2011).

Nutrient storage tissues: Surrounding the embryo are nutrient storage tissues, which provide the energy reserves needed to support germination and initial growth of the young plant. These reserves can be stored in the form of starch, proteins, lipids or other organic compounds (Barthet & Daun, 2011).

Protective shell: The seed is encased in a protective shell, usually formed by the ovule's envelopes. This shell protects the embryo and nutrient storage tissues from mechanical damage, infection and desiccation during storage and dispersal (Barthet & Daun, 2011).

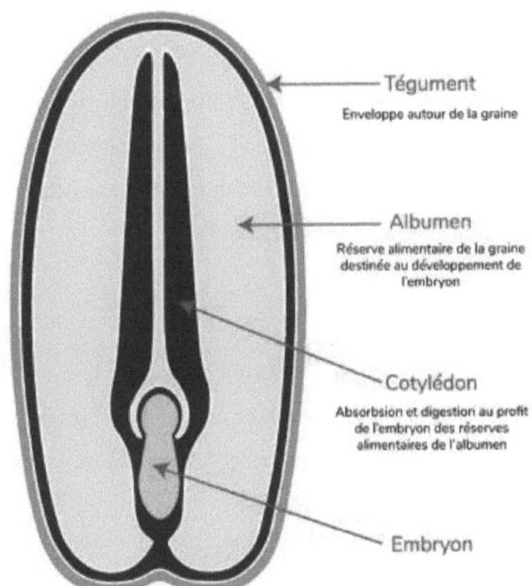

Figure 23. Seed morphology.
(*The Secret of Seeds*, n. d.)

3.1.4. Seed function

The seed is an essential structure for the dispersal and survival of plant species. It contains all the genetic information and resources needed for the new plant to develop and grow after germination. The seed may remain dormant during periods of unfavorable conditions, then germinate when conditions are conducive to growth (Leopold et al., 1994).

3.1.5. Seed dispersal

Once the seeds are fully developed, the ripe fruit cracks or opens to release the seeds. Seeds can be dispersed by a variety of mechanisms, such as wind, water, animals, birds or humans. This dispersal enables plants to colonize new habitats and reproduce successfully in a variety of environments (Leopold et al., 1994) .

Seed formation is an essential stage in the life cycle of flowering plants, ensuring the propagation and survival of plant species in their environment. The complex structure of the seed and its dispersal mechanisms bear witness to the evolutionary adaptation of plants to optimize reproduction and survival in a wide range of environmental conditions (Leopold et al., 1994)..

3.2. Fruit development

Fruit development is an essential phase in the reproductive process of flowering plants (angiosperms), following fertilization and seed formation. This stage marks the transformation of the fertilized ovary into a mature structure called a fruit. Here's a detailed exploration of fruit development (Kigel, 1995) :

3.2.1. Fruit origin

Fruits develop from the ovary of the fertilized flower. The ovary may be simple, originating from a single carpel (simple fruit), or compound, formed from several fused carpels (compound fruit). This distinction influences the structure and composition of the final fruit (Kigel, 1995)..

3.2.2. Fruit diversity

Fruits come in a wide variety of shapes, colors, textures and chemical compositions, adapted to their mode of dispersal and environment. They can be classified into different categories according to their structure, such as berries, drupes, capsules, pods, achenes, cones, samaras, etc. (Kigel, 1995).

3.2.3. Fruit characteristics

The fruits have a variety of adaptive characteristics to facilitate their dispersal:

Some fruits are fleshy and sweet, making them attractive to animals, which eat the flesh and disperse the intact seeds in their environment. These fruits are often colored to attract animals visually and are rich in nutrients to reward them (Kigel, 1995).

Other fruits are dry and dehiscent, meaning they open spontaneously at maturity to release the seeds. These fruits can be dispersed by wind (anemochory), water (hydrochory) or other mechanical means (Kigel, 1995).

3.2.4. Importance of Dispersion

Fruit and seed dispersal is essential for the colonization of new habitats and the survival of plants through the next generation. It enables the spatial dispersal of individuals, reduces intra-specific competition and promotes genetic diversity and the colonization of new territories.

3.2.5. Interactions with Animals

Fruits play a crucial role in plant-animal interactions, providing a food source for many organisms, such as birds, mammals, reptiles and insects. These interactions can benefit the plant by promoting seed dispersal and enabling cross-pollination.

Fruit development is a complex and highly regulated process that reflects the evolutionary adaptation of plants to their environment and to their interactions with other organisms. The diversity of fruits and their dispersal strategies testifies to the richness and complexity of plant life(Kigel, 1995).

F. Communication in the Plant World

1. Chemical and electrical signals between plants

Communication between plants via chemical and electrical signals is a fast-growing area of research in plant biology. This complex process reveals an intriguing dynamic within plant communities, where plants are able to detect and respond to a diverse range of environmental stimuli (García-Servín et al., 2021).

1.1. Diversity of Chemical Signals

The diversity of chemical signals emitted by plants is remarkable, testifying to their ability to adapt and respond in specific ways to environmental threats. This variety of chemical signals enables plants to communicate effectively with their environment and modulate their responses according to the type and intensity of the perceived threat. Here are a few examples illustrating this diversity of chemical signals:

i. Specific pheromones

Some plants emit specific pheromones in response to insect attack. These pheromones act as alarm signals, alerting neighboring plants to the presence of potential predators. For example, when a plant is attacked by caterpillars, it may release specific pheromones that attract the caterpillars' natural predators, such as parasitoid wasps, which help reduce the insect pest population (García-Servín et al., 2021).

ii. Secondary toxic metabolites

Other plants produce toxic secondary metabolites in response to herbivore attack. These metabolites can act as chemical defense agents, deterring herbivores from feeding or making them sick if they ingest plant tissue. For example, some plants produce compounds such as alkaloids, terpenes or glucosinolates, which have repellent or toxic effects on herbivores and can reduce the damage caused by predation (García-Servín et al., 2021).

iii. Volatile Defense

In response to an attack by pathogens or insects, some plants emit volatile defense compounds. These volatile compounds act as alarm signals for neighboring plants, triggering similar defense mechanisms. For example, when a plant is infected by a pathogenic fungus, it may release volatile compounds that activate defense mechanisms in surrounding plants, increasing their resistance to disease (García-Servín et al., 2021).

iv. Hormonal signaling

Plants also use plant hormones, such as jasmonic acid and salicylic acid, to regulate their responses to environmental stresses. These hormones act as internal chemical signals, coordinating the implementation of adaptive responses, such as the production of defense metabolites or the closure of stomata to reduce water loss through transpiration (García-Servín et al., 2021).

Together, this diversity of chemical signals enables plants to modulate their responses according to the specific threats they face in their environment. This capacity for adaptation and chemical communication is essential for plants to survive and thrive in often hostile and competitive environments. By understanding these complex chemical communication mechanisms, scientists can better understand how plants interact with their environment and develop more effective management strategies to promote the health and resilience of plant ecosystems (Fromm & Lautner, 2007).

1.2. Behavioral responses

The behavioral responses of plants to the chemical signals emitted by their conspecifics are remarkable, demonstrating their ability to perceive and react adaptively to their environment. These behavioral responses can take many forms, and are often coordinated to maximize chances of survival and reproduction in dynamic, competitive environments. Here are a few examples of these complex behavioral responses:

i. Growth Rate Adjustment

In response to chemical distress signals from other plants, some species can adjust their growth rate to optimize their allocation of resources. For example, when neighboring plants are subject to intense competition for light, nutrients or water, a plant may slow its growth in order to reduce competition and optimize the use of available resources (Fromm & Lautner, 2007).

ii. Resource allowance

Plants can also adjust their resource allocation in response to chemical signals of distress. For example, when a plant detects the presence of chemical signals indicating herbivore attack, it may reallocate its resources towards the production of defensive metabolites, such as toxic compounds or volatile compounds, to deter herbivores or limit damage caused by predation (Fromm & Lautner, 2007).

iii.   Metabolite production

In response to chemical signals of distress, plants can also increase their production of specialized metabolites involved in defense against herbivores, pathogens or other environmental stresses. For example, some plants may increase their production of antioxidants or enzymes involved in the detoxification of harmful substances, in order to protect their tissues against oxidative damage or pathogenic attack (Fromm & Lautner, 2007).

iv.   Root System

Behavioral responses are not limited to the aerial parts of plants, but can also affect the root system. For example, in response to chemical signals indicating competition for nutrients or water in the soil, plants may modify the growth or branching of their roots to exploit available resources more efficiently and minimize competition with neighbors (Fromm & Lautner, 2007).

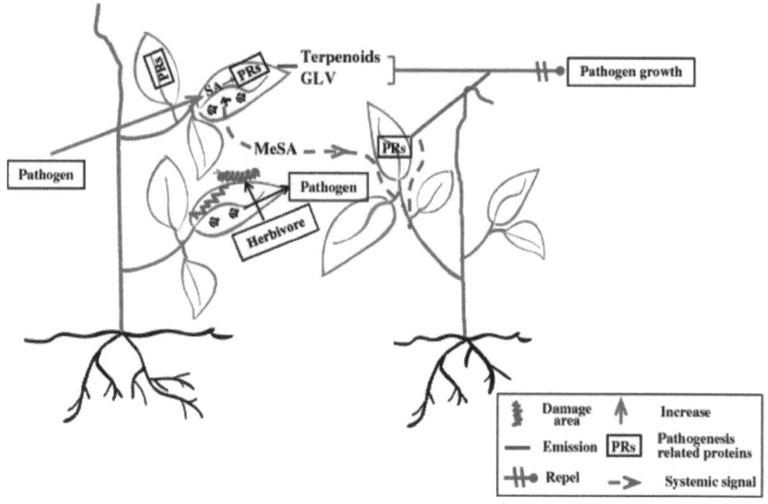

These examples illustrate the ability of plants to perceive and respond adaptively to chemical signals emitted by their conspecifics. This behavioral plasticity is essential for maximizing chances of survival and reproduction in often competitive and changing environments. By understanding these complex behavioral response mechanisms, scientists can better understand how plants interact with their environment and develop more effective management strategies to promote the health and resilience of plant ecosystems (Volkov et al., 2019).

v.   Interactions with soil microorganisms

Chemical signals exchanged between plants can also influence interactions with soil microorganisms, such as bacteria and fungi. For example, plant roots can release specific organic compounds to attract beneficial microorganisms that promote plant health, such as nitrogen-fixing rhizobia or nutrient-absorbing mycorrhizae (Enagbonma et al., 2023).

1.3. Electrical communication

Electrical communication between plants is a fascinating and relatively recently discovered mechanism that complements chemical communication and helps coordinate plants' adaptive responses to environmental stimuli. These electrical signals, often triggered by events such as physical damage or predator attack, propagate through plant tissues at surprising speeds, enabling rapid coordination of responses throughout the plant. Here are some key points about electrical communication between plants:

i.   Rapid propagation

Electrical signals can propagate through plant tissues at speeds of up to several centimetres per second, enabling rapid communication between different parts of the plant. This rapid propagation is essential for coordinating adaptive responses throughout the plant in the event of stress or threat (Volkov et al., 2019).

ii.   Coordination of Adaptive Responses

When one part of the plant is damaged, electrical signals can propagate to other parts of the plant to trigger defense mechanisms. For example, a damaged leaf may emit electrical signals that propagate to the roots, triggering an increase in the production of defensive chemical compounds or the closure of stomata to reduce water loss (Volkov et al., 2019).

iii.   Role in Defense and Repair

Electrical signals play a crucial role in coordinating defense and repair responses in plants. In addition to triggering immediate defense mechanisms, such as the production of toxic chemical compounds, electrical signals can also stimulate cell growth and wound healing to repair damage caused by environmental stresses (Volkov et al., 2019).

iv.   Interaction with other signals

Electrical communication between plants can interact with other signaling systems, such as chemical and hormonal signals, to coordinate complex adaptive responses. This integration of different modes of communication enables fine, precise regulation of plant responses to environmental stimuli (Volkov et al., 2019).

By understanding these complex communication mechanisms between plants, researchers can better understand how plants perceive and respond to their environment. This knowledge can have practical applications in fields such as agriculture, forestry and biodiversity conservation, helping to develop ecosystem management strategies that promote the health and resilience of plant communities. Ultimately, plant communication reveals the remarkable adaptability and complexity of plant organisms in their quest to survive and thrive in ever-changing environments.

2. Interaction with other organisms, such as insects and fungi

The interaction of plants with other organisms, notably insects and fungi, is a crucial aspect of plant biology, profoundly influencing plant health, growth and survival, as well as the structure and dynamics of ecosystems. This complex, multifaceted interaction between plants and other organisms is the result of millions of years of coevolution, in which plants have developed a series of defense, seduction and cooperation mechanisms to adapt to their environment and ensure their reproductive success (Prisa, 2023). Here's a detailed exploration of how plants interact with insects and fungi:

2.1. Interaction with insects

Insects play a major role in plant life, acting both as essential pollinators for the sexual reproduction of flowering plants and as voracious herbivores that can cause considerable damage to crops and natural ecosystems. This interaction between plants and insects is often marked by dynamic coevolution, with plants developing a variety of defense mechanisms to deter herbivores and attract pollinators. These defense mechanisms include the production of toxic or repellent chemical compounds, the presence of physical structures such as spines or hairs, and symbiotic partnerships with predatory or parasitoid insects that act as biological pest control agents.

Plants have also evolved to produce a diverse range of chemical and visual signals to attract pollinating insects and ensure their reproduction. Flowers often produce bright colors, sweet scents and nectar to attract insects, while host plants provide habitat and food resources for insect larvae. This complex interaction between plants and insects influences not only the structure and diversity of plant communities, but also the overall biodiversity of terrestrial ecosystems.

2.2. Interaction with Mushrooms

Fungi also play a crucial role in plant life, acting as symbiotic partners in mutualistic associations such as mycorrhizae, or as pathogens causing fungal diseases (da Silva Folli-Pereira et al., 2013). Mycorrhizae are symbiotic associations between plant roots and fungal hyphae, which facilitate the uptake of soil nutrients, such as nitrogen and phosphorus, in exchange for organic compounds produced by the plant (da Silva Folli-Pereira et al., 2013). This mutualistic symbiosis is widespread in the plant kingdom and plays a crucial role in plant nutrition, growth and stress resistance, as well as in ecosystem dynamics.

However, not all fungi are beneficial to plants, as some fungi can cause fungal diseases that compromise crop health and productivity. Fungal diseases can result in damage to plant tissue, reduced growth and yield, and even plant death in severe cases (da Silva Folli-Pereira et al., 2013).. Plants have developed a series of defense mechanisms to protect themselves against fungal diseases, including the production of antifungal chemical compounds, the formation of physical barriers such as cell wall thickness, and immune responses induced by specific molecular signals.

In summary, the interaction of plants with insects and fungi is a fundamental aspect of plant biology, shaping the structure and dynamics of terrestrial ecosystems. This complex interaction between plants and other organisms is the result of ongoing coevolution, in which plants have developed a series of defense, seduction and cooperation mechanisms to adapt to their environment and ensure their reproductive success. By understanding these interaction mechanisms, scientists can better understand how plants interact with their environment and develop more effective management strategies to promote the health and resilience of plant ecosystems. (da Silva Folli-Pereira et al., 2013).

3. Impact of environmental factors on plant communication

The impact of environmental factors on plant communication is an exciting area of research that explores how environmental conditions influence the production, transmission and reception of chemical and electrical signals between plants. These complex interactions between plants and their environment can modulate plant communication in a variety of ways, influencing plant survival, growth and reproduction, as well as the structure and dynamics of ecosystems (da Silva Folli-Pereira et al., 2013). Here is an in-depth analysis of the impact of environmental factors on plant communication:

3.1. Light

Light is one of the main environmental factors influencing plant communication. The availability of light affects the production of volatile chemical compounds in leaves, which are often involved in communication between plants. For example, plants exposed to intense light may produce more volatile compounds to signal the presence of predators or to attract pollinators. In addition, light also influences electrical communication between plants, as electrical signals can be triggered by rapid changes in light intensity or by interactions with predators (Komine & Nakagawa, 2003).

3.2. Temperature

Temperature also plays a crucial role in plant communication. Plants can adjust their production of chemical signals in response to temperature fluctuations to optimize their efficiency under changing environmental conditions. For example, some plants produce more volatile compounds in response to higher temperatures to signal heat damage or to attract pollinators adapted to warmer climates. In addition, temperature affects the propagation speed of electrical signals in plant tissues, which can influence the coordination of adaptive responses (Ding & Yang, 2022; Morison & Lawlor, 1999).

3.3. Humidity

Atmospheric and soil humidity can also influence plant communication. Plants can adjust their production of chemical signals in response to high or low humidity levels to signal stressful environmental conditions such as drought or excessive moisture. In addition, soil moisture can influence the production of electrical signals in plant roots, which can affect the coordination of adaptive responses such as the regulation of water and nutrient uptake (Baluška et al., 2006).

3.4. Soil composition

Soil composition, including the availability of nutrients, minerals and organic compounds, can also influence plant communication. Plants can adjust their production of chemical signals in response to variations in soil composition to optimize growth and development. In addition, the presence of beneficial or pathogenic microorganisms in the soil can alter the production of chemical and electrical signals by plants, which can have consequences for their interaction with other soil organisms and for ecosystem health (Davies et al., 1994).

In summary, the impact of environmental factors on plant communication is a crucial aspect of plant biology, influencing plant survival, growth and reproduction, as well as ecosystem structure and dynamics. By understanding these complex interactions, scientists can better understand how plants interact with their environment and develop more effective management strategies to promote the health and resilience of plant ecosystems.

G. Adapting to environmental stress

Plants face a variety of environmental stresses, from extreme climatic conditions such as heat, cold and drought, to attacks by microbial pathogens. This chapter explores the physiological responses of plants to these abiotic and biotic stresses, as well as the adaptive mechanisms that enable them to survive and thrive in often hostile and changing environments.

1. Physiological Responses to Abiotic Stress

Abiotic stresses, such as heat, cold, drought, flooding and saline soils, can significantly disrupt plant metabolism and growth. In this context, plants have developed a range of physiological mechanisms to cope with these stresses and maintain their homeostasis. For example, in the event of excessive heat, plants can activate thermotolerance mechanisms, such as the synthesis of heat shock proteins, which protect cellular structures against heat damage. Similarly, in response to drought, plants can regulate transpiration and stomatal activity to limit water loss and maintain internal hydration (Harfouche et al., 2014).

The physiological responses of plants to abiotic stresses are complex and dynamic mechanisms that enable them to cope with adverse environmental conditions such as heat, cold, drought, flooding and saline soils. These physiological responses involve a series of biochemical, molecular and cellular modifications aimed at maintaining internal homeostasis, protecting cellular structures and ensuring overall plant survival (Harfouche et al., 2014). Here's a more detailed exploration of these physiological responses:

1.1. Sweat and Ostomy Regulation

When plants are confronted with hot or dry conditions, one of the first physiological responses is the regulation of transpiration and stomatal opening. Stomata, located mainly on leaf surfaces, control water loss through transpiration and also regulate gas exchange, such as photosynthesis and respiration. In response to conditions of drought or excessive heat, plants may partially or totally close their stomata to limit water loss through transpiration, thereby preserving their internal hydration and cellular integrity (Yoo et al., 2009).

1.2. Thermal Shock Protein Synthesis

Heat shock proteins (HSPs) are a group of highly conserved proteins that play an essential role in protecting cells against thermal damage. Under heat stress, plants activate the synthesis of these proteins, which act as molecular chaperones, facilitating correct protein folding and protecting cellular structures from damage. HSPs also help stabilize cell membranes and protect heat-sensitive enzyme complexes (Krishnan et al., 1989, p. 1).

1.3. Protective Metabolite Accumulation

Plants can accumulate various protective metabolites, such as osmoprotectants and antioxidants, in response to abiotic stresses (Harborne, 2007). Osmoprotectants, such as proline and soluble sugars, act by maintaining osmotic homeostasis and protecting proteins and cell membranes from damage caused by dehydration. Antioxidants, such as flavonoids and carotenoids, act by neutralizing reactive oxygen species (ROS) produced in excess during stressful conditions, thus protecting cells against oxidative stress (Laoué et al., 2022).

1.4. Reorganization of Cell Structure

In response to abiotic stresses, plants can reorganize their cell structure to optimize their adaptation to changing environmental conditions. For example, some plants may alter the composition of their cell membranes to maintain fluidity at extreme temperatures, while others may increase the density of their vascular tissues to improve water and nutrient transport (Conde et al., 2011). In addition, plants can induce the formation of healing tissues to repair damage caused by mechanical or environmental stresses (Chaudhry & Sidhu, 2022).

In summary, the physiological responses of plants to abiotic stresses are complex adaptations that enable them to survive and thrive in often hostile environments. By understanding the mechanisms underlying these responses, scientists can develop strategies to improve crop resilience in the face of increasing environmental stresses, thereby helping to ensure global food security and the sustainability of terrestrial ecosystems.

2. Defense mechanisms against pathogens

Plant defense mechanisms against microbial pathogens are a fascinating and crucial area of research in plant biology. Faced with attacks from fungi, bacteria and viruses, plants have evolved a series of sophisticated strategies to detect, combat and neutralize pathogenic invaders (Garcion et al., 2014). These defense mechanisms are essential to ensure the survival and health of plants in often hostile and competitive environments. In this section, we will explore in detail the different components of plant defense mechanisms against pathogens, highlighting their complexity and importance in plant biology (Garcion et al., 2014).

2.1. Recognition of Pathogen-Associated Molecular Patterns (PAMPs)

One of the first steps in the plant immune response to pathogens is the recognition of pathogen-associated molecular patterns (PAMPs) by plant receptors (Postel & Kemmerling, 2009). PAMPs are conserved molecular motifs present in pathogen cells, such as cell wall polysaccharides or surface proteins, which are recognized by plant receptors, known as pattern recognition receptors (PRRs). When a PAMP binds to a PRR, it triggers a cascade of biochemical signals that lead to the activation of the plant's defense mechanisms (Rathore & Ghosh, 2018).

2.2. Immune Responses

Following recognition of PAMPs, plants activate various immune responses to counter pathogen infection. These immune responses include the production of antimicrobial molecules such as phytoalexins, antimicrobial proteins and cell wall-degrading enzymes. Phytoalexins are secondary metabolites synthesized in response to pathogen infection, and act by inhibiting pathogen growth. Antimicrobial proteins, such as chitinases and glucanases, degrade the cellular components of pathogens, while cell wall-degrading enzymes, such as pectinases and cellulases, break down the cell walls of pathogens to neutralize them (Nürnberger & Kemmerling, 2008).

2.3. Activate signal channels

Plant immune responses are mediated by complex signaling pathways that transmit molecular signals from the pathogen recognition site throughout the plant (Parker, 2000). These pathways involve a variety of signaling molecules, such as plant hormones, kinases and transcription factors, which coordinate plant immune responses. For example, salicylic acid (SA) is a plant hormone involved in the plant immune response against biotrophic pathogens, while jasmonic acid (JA) and ethylene (ET) are involved in the immune response against necrotrophic pathogens (Liu et al., 2016).

2.4. Strengthening Plant Resistance

By activating these defense mechanisms, plants strengthen their resistance to pathogens and limit their ability to colonize and spread through plant tissues. This reinforcement of plant resistance is crucial for limiting disease damage and ensuring plant survival and health in often hostile environments. In addition, plant defense mechanisms can also induce systemic responses that strengthen plant resistance to future pathogen infections.

In summary, plant defense mechanisms against microbial pathogens are complex and dynamic processes involving recognition of PAMPs, activation of immune responses, molecular signaling and enhancement of plant resistance. These mechanisms are essential for ensuring plant survival and health in often hostile and competitive environments, and represent a crucial area of research in plant biology and crop protection.

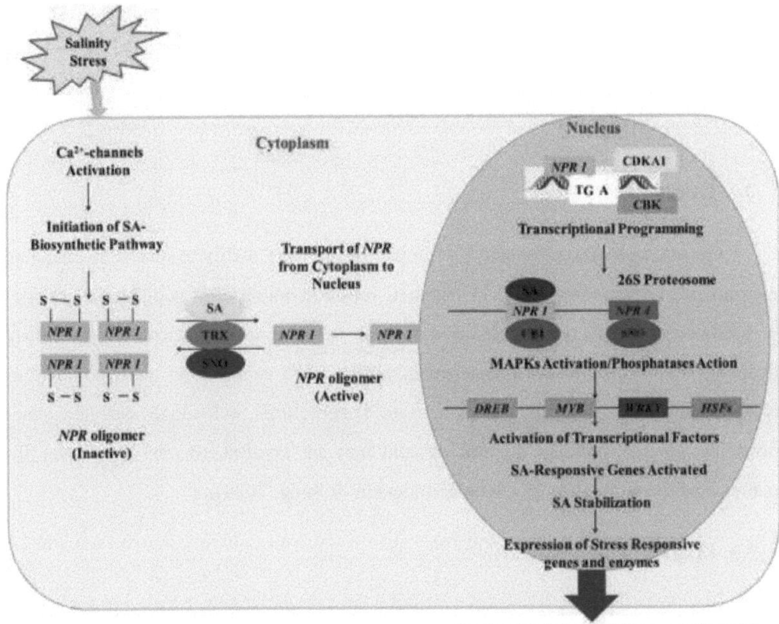

Figure 25. Overview of salicylic acid signaling in salt-stressed plants and the mechanism of induced stress tolerance.
(Sharma et al., 2023)

3. Importance of Resilience and Adaptation

In changing and unpredictable environments, resilience and adaptation are essential for the long-term survival of plants. Plants display remarkable phenotypic plasticity, enabling them to modulate their physiology, morphology and behavior in response to changing environmental conditions. This phenotypic plasticity is the result of a combination of genetic and epigenetic responses that enable plants to adjust rapidly to stress and maximize their reproductive success in variable environments.

3.1. Phenotypic plasticity

The phenotypic plasticity of plants is a fundamental mechanism in their adaptation to environmental change. This plasticity manifests itself in the ability of plants to adjust their physiology, morphology and behavior to changing environmental conditions (Nicotra et al., 2010). For example, faced with reduced water availability, plants may develop deeper roots to exploit underground water reserves, or reduce their transpiration rate to limit water loss. Similarly, in response to extreme temperatures, plants may modify their metabolism to maximize their energy efficiency and tolerance to heat or cold (Nicotra et al., 2010).

3.2. Genetic adaptation

Genetic adaptation also plays a crucial role in plants' ability to survive in changing environments (Anderson et al., 2011). Genetic variation within plant populations provides the substrate on which natural selection can act, favoring traits that confer an adaptive advantage under specific environmental conditions. For example, in plant populations subject to selection pressures due to extreme temperatures, individuals carrying genes conferring greater tolerance to heat or cold may be favored, thereby increasing the resilience of the population as a whole(Anderson & Song, 2020).

3.3. Epigenetic regulation

In addition to genetic variation, epigenetic mechanisms enable plants to regulate gene expression in response to environmental signals (Guarino et al., 2022). These epigenetic mechanisms, such as DNA methylation and histone modifications, can modulate gene accessibility and influence gene expression without altering the DNA sequence. Thus, plants can adapt their phenotype to changing environmental conditions by regulating the expression of genes involved in processes such as stress response, growth and development (Ashapkin et al., 2020).

Conclusion

Advanced Plant Physiology allowed us to delve into the complex mechanisms that govern plant life, from root to flower. We explored in detail how plants absorb nutrients from the soil, carry out photosynthesis and cellular respiration, and interact with their environment to ensure growth and survival. From leaf anatomy to chemical and electrical communication between plants, from adaptations to environmental stresses to defense mechanisms against pathogens, every aspect of plant physiology has been carefully examined.

However, our exploration must not be limited to the pages of this book. Applying this knowledge is crucial to optimizing plant growth, improving global food security and fostering sustainable development. Whether in agriculture, forestry, biotechnology or ecology, a thorough understanding of plant physiology can pave the way for more efficient and environmentally-friendly innovations and practices.

As we conclude this journey through plant physiology, we need to keep an eye on the future. New technologies are emerging, new questions are being asked and new challenges are presenting themselves. Plant physiology research will continue to evolve to meet the changing needs of our planet and its growing population. By investing in basic and applied research, encouraging collaboration and remaining open to innovative ideas, we can continue to advance our understanding of plant life and meet the environmental challenges we face.

References

Anatomy of a LEAF + explanatory DIAGRAM. (n. d.). projetecolo.com. Retrieved February 11, 2024, from https://www.projetecolo.com/anatomie-d-une-feuille-schema-d-une-feuille-11.html

Anderson, J. T., & Song, B.-H. (2020). Plant adaptation to climate change-Where are we? *Journal of Systematics and Evolution,* *58*(5), 533-545. https://doi.org/10.1111/jse.12649

Anderson, J. T., Willis, J. H., & Mitchell-Olds, T. (2011). Evolutionary genetics of plant adaptation. *Trends in Genetics,* *27*(7), 258-266. https://doi.org/10.1016/j.tig.2011.04.001

Ashapkin, V. V., Kutueva, L. I., Aleksandrushkina, N. I., & Vanyushin, B. F. (2020). Epigenetic Mechanisms of Plant Adaptation to Biotic and Abiotic Stresses. *International Journal of Molecular Sciences,* *21*(20), Article 20. https://doi.org/10.3390/ijms21207457

Baluška, F., Mancuso, S., & Volkmann, D. (Eds.). (2006). *Communication in Plants.* Springer. https://doi.org/10.1007/978-3-540-28516-8

Barber, S., & Silberbush, M. (1984). Plant root morphology and nutrient uptake. *Roots, nutrient and water influx, and plant growth, 49,* 65-87.

Barthet, V. J., & Daun, J. K. (2011). 5-Seed morphology, composition, and quality. In J. K. Daun, N. A. M. Eskin, & D. Hickling (Eds.), *Canola* (pp. 119-162). AOCS Press. https://doi.org/10.1016/B978-0-9818936-5-5.50009-7

Belhadi, T., & Zillal, C. (2020). *Literature review on the involvement of Fluorescent Rhizobacteris Pseudomonas. Fluorescent Spp in plant growth enhancement.*

Berry, J., & Bjorkman, O. (1980). Photosynthetic Response and Adaptation to Temperature in Higher Plants. *Annual Review of Plant Physiology, 31*(1), 491-543. https://doi.org/10.1146/annurev.pp.31.060180.002423

Boussac, A., & Mathis, P. (2015). 17. Water oxidation at the heart of photosynthesis. In A. Euzen, C. Jeandel, & R. Mosseri (Eds.), *L'eau à découvert* (pp. 80-81). CNRS Éditions. https://doi.org/10.4000/books.editionscnrs.9824

Brett, C. T., & Waldron, K. W. (1996). *Physiology and biochemistry of plant cell walls* (Vol. 2). Springer Science & Business Media.

Cellular respiration-Definition and Examples-Biology Online Dictionary (2023, June 14). Biology Articles, Tutorials & Dictionary Online. https://www.biologyonline.com/dictionary/cellular-respiration

Chaudhry, S., & Sidhu, G. P. S. (2022). Climate change regulated abiotic stress mechanisms in plants: A comprehensive review. *Plant Cell Reports, 41*(1), 1-31. https://doi.org/10.1007/s00299-021-02759-5

Collin, P. (2001). Environmental adaptation in vascular plants. *L'Année Biologique, 40,* 21-42. https://doi.org/10.1016/S0003-5017(01)72083-1

Conde, A., Chaves, M. M., & Gerós, H. (2011). Membrane Transport, Sensing and Signaling in Plant Adaptation to Environmental Stress. *Plant and Cell Physiology, 52*(9), 1583-1602. https://doi.org/10.1093/pcp/pcr107

Cruiziat, P., Améglio, T., & Cochard, H. (2001). Cavitation: A mechanism disrupting water circulation in plants. *Mécanique & Industries, 2*(4), 289-298. https://www.sciencedirect.com/science/article/pii/S1296213901011101

Das, A., Lee, S.-H., Hyun, T. K., Kim, S.-W., & Kim, J.-Y. (2013). Plant volatiles as method of communication. *Plant Biotechnology Reports, 7*(1), 9-26. https://doi.org/10.1007/s11816-012-0236-1

da Silva Folli-Pereira, M., Sant'Anna Meira-Haddad, L., Vilhena da Cruz Houghton, C. M. N. S. de, & Megumi Kasuya, M. C. (2013). Plant-Microorganism Interactions: Effects on the Tolerance of Plants to Biotic and Abiotic Stresses. In K. R. Hakeem, P. Ahmad, & M. Ozturk (Eds.), *Crop Improvement: New Approaches and Modern Techniques* (pp. 209-238). Springer US. https://doi.org/10.1007/978-1-4614-7028-1_6

Davies, W. J., Tardieu, F., & Trejo, C. L. (1994). How Do Chemical Signals Work in Plants that Grow in Drying Soil? *Plant Physiology, 104*(2), 309-314. https://www.ncbi.nlm.nih.gov/pmc/articles/PMC159200/

Delaporte, D. (2009). Botany: Chemical constituents of the root and other parts of the plant in relation to the terrain. In P. Goetz, D. Delaporte, & P. Stoltz (Eds.), *Le*

Ginseng : Vertus thérapeutiques d'une plante adaptogène (pp. 31-59). Springer. https://doi.org/10.1007/978-2-287-79924-2_3

Dennis Jr, F. G. (1984). Fruit Development. In *Physiological Basis of Crop Growth and Development* (pp. 265-289). John Wiley & Sons, Ltd. https://doi.org/10.2135/1984.physiologicalbasis.c10

Ding, Y., & Yang, S. (2022). Surviving and thriving: How plants perceive and respond to temperature stress. *Developmental Cell, 57*(8), 947-958. https://doi.org/10.1016/j.devcel.2022.03.010

Dumas, C. (2001). Reproduction and development of flowering plants. *Comptes Rendus de l'Académie des Sciences - Series III - Sciences de la Vie, 324*(6), 517-521. https://doi.org/10.1016/S0764-4469(01)01320-8

El Mekdad, F. (2023). *Can rhizodeposition in deep soil horizons store carbon?*

Enagbonma, B. J., Fadiji, A. E., Ayangbenro, A. S., & Babalola, O. O. (2023). Communication between Plants and Rhizosphere Microbiome: Exploring the Root Microbiome for Sustainable Agriculture. *Microorganisms, 11*(8), Article 8. https://doi.org/10.3390/microorganisms11082003

Evert, R. F. (2006). *Esau's plant anatomy: Meristems, cells, and tissues of the plant body: Their structure, function, and development.* John Wiley & Sons.

Ferry, J. F., & Ward, H. S. (1959). *Fundamentals of plant physiology.*

Nitrogen fixation: Definition and explanations. (n.d.). AquaPortail. Retrieved February 11, 2024, from https://www.aquaportail.com/dictionnaire/definition/2236/fixation-de-l-azote

Flower: Definition, anatomy, role, photos, uses... (s. d.). AquaPortail. Retrieved February 12, 2024, from https://www.aquaportail.com/dictionnaire/definition/7708/fleur

Fortin, J. A., Plenchette, C., & Piché, Y. (2008). Mycorrhizae. *The new green revolution. MultiMonde Quae.(Eds.), Quebec.*

Friedman, J., & Barrett, S. C. H. (2009). Wind of change: New insights on the ecology and evolution of pollination and mating in wind-pollinated plants. *Annals of Botany, 103*(9), 1515-1527. https://doi.org/10.1093/aob/mcp035

Fromm, J., & Lautner, S. (2007). Electrical signals and their physiological significance in plants. *Plant, Cell & Environment*, *30*(3), 249-257. https://doi.org/10.1111/j.1365-3040.2006.01614.x

Futura (s. d.). *Definition | Calvin Cycle-Cycle de Calvin-Benson | Futura Planète*. Futura. Retrieved February 11, 2024, from https://www.futura-sciences.com/planete/definitions/botanique-cycle-calvin-7427/

Galiana, A. (1990). *Nitrogen-fixing symbiosis in Acacia mangium-rhizobium.*

García-Servín, M. Á., Mendoza-Sánchez, M., & Contreras-Medina, L. M. (2021). Electrical signals as an option of communication with plants: A review. *Theoretical and Experimental Plant Physiology*, *33*(2), 125-139. https://doi.org/10.1007/s40626-021-00203-3

Garcion, C., Lamotte, O., Cacas, J.-L., & Métraux, J.-P. (2014). Mechanisms of Defence to Pathogens: Biochemistry and Physiology. In *Induced Resistance for Plant Defense* (pp. 106-136). John Wiley & Sons, Ltd. https://doi.org/10.1002/9781118371848.ch6

Givnish, T. (1979). On the Adaptive Significance of Leaf Form. In O. T. Solbrig, S. Jain, G. B. Johnson, & P. H. Raven (Eds.), *Topics in Plant Population Biology* (pp. 375-407). Macmillan Education UK. https://doi.org/10.1007/978-1-349-04627-0_17

Guarino, F., Cicatelli, A., Castiglione, S., Agius, D. R., Orhun, G. E., Fragkostefanakis, S., Leclercq, J., Dobránszki, J., Kaiserli, E., Lieberman-Lazarovich, M., Sõmera, M., Sarmiento, C., Vettori, C., Paffetti, D., Poma, A. M. G., Moschou, P. N., Gašparović, M., Yousefi, S., Vergata, C., ... Martinelli, F. (2022). An Epigenetic Alphabet of Crop Adaptation to Climate Change. *Frontiers in Genetics*, *13*. https://www.frontiersin.org/journals/genetics/articles/10.3389/fgene.2022.81872 7

Guo, L., Luo, X., Li, M., Joldersma, D., Plunkert, M., & Liu, Z. (2022). Mechanism of fertilization-induced auxin synthesis in the endosperm for seed and fruit development. *Nature Communications*, *13*(1), Article 1. https://doi.org/10.1038/s41467-022-31656-y

Gurrieri, L., Fermani, S., Zaffagnini, M., Sparla, F., & Trost, P. (2021). Calvin-Benson cycle regulation is getting complex. *Trends in Plant Science, 26*(9), 898-912. https://doi.org/10.1016/j.tplants.2021.03.008

Harborne, J. B. (2007). Role of Secondary Metabolites in Chemical Defence Mechanisms in Plants. In *Ciba Foundation Symposium 154-Bioactive Compounds from Plants* (pp. 126-139). John Wiley & Sons, Ltd. https://doi.org/10.1002/9780470514009.ch10

Harfouche, A., Meilan, R., & Altman, A. (2014). Molecular and physiological responses to abiotic stress in forest trees and their relevance to tree improvement. *Tree Physiology, 34*(11), 1181-1198. https://doi.org/10.1093/treephys/tpu012

Hermann, K., & Kuhlemeier, C. (2011). The genetic architecture of natural variation in flower morphology. *Current Opinion in Plant Biology, 14*(1), 60-65. https://doi.org/10.1016/j.pbi.2010.09.012

Hodge, A., Berta, G., Doussan, C., Merchan, F., & Crespi, M. (2009). *Plant root growth, architecture and function.*

Hopkins, W. G. (2003). *Plant physiology.* De Boeck Supérieur.

Jain, V. (2018). *Fundamentals of plant physiology.* S. Chand Publishing.

Kevan, P. G., & Baker, H. G. (1983). Insects as Flower Visitors and Pollinators. *Annual Review of Entomology, 28*(1), 407-453. https://doi.org/10.1146/annurev.en.28.010183.002203

Kigel, J. (1995). *Seed Development and Germination.* CRC Press.

Kolb, E., Legué, V., & Bogeat-Triboulot, M.-B. (2017). Physical root-soil interactions. *Physical biology, 14*(6), 065004.

Komine, T., & Nakagawa, M. (2003). Integrated system of white LED visible-light communication and power-line communication. *IEEE Transactions on Consumer Electronics, 49*(1), 71-79. https://doi.org/10.1109/TCE.2003.1205458

Krishnan, M., Nguyen, H. T., & Burke, J. J. (1989). Heat Shock Protein Synthesis and Thermal Tolerance in Wheat 1. *Plant Physiology, 90*(1), 140-145. https://doi.org/10.1104/pp.90.1.140

Photosynthesis: A subject for future study. (2020, July 22). Oleomac blog. https://blog.oleomac.fr/la-photosynthese/

Laoué, J., Fernandez, C., & Ormeño, E. (2022). Plant Flavonoids in Mediterranean Species: A Focus on Flavonols as Protective Metabolites under Climate Stress. *Plants, 11*(2), Article 2. https://doi.org/10.3390/plants11020172

The Secret of Seeds (n.d.). Retrieved February 12, 2024, from http://www.laviesaine.fr/actualites/lepointsur-/1799/lesecretdesgraines/

The root system. (s. d.). Biology101. Retrieved February 11, 2024, from https://www.biologie101.fr/biologie-vegetale/appareil-vegetatif-plante/systeme-racinaire

Leopold, A. C., Sun, W. Q., & Bernal-Lugo, I. (1994). The glassy state in seeds: Analysis and function. *Seed Science Research, 4*(3), 267-274. https://doi.org/10.1017/S0960258500002294

Liu, L., Sonbol, F.-M., Huot, B., Gu, Y., Withers, J., Mwimba, M., Yao, J., He, S. Y., & Dong, X. (2016). Salicylic acid receptors activate jasmonic acid signalling through a non-canonical pathway to promote effector-triggered immunity. *Nature Communications, 7*(1), Article 1. https://doi.org/10.1038/ncomms13099

Martin, A. (2022, June 9). Sexual reproduction in flowering plants. *Online Learning College.* https://online-learning-college.com/knowledge-hub/gcses/gcse-biology-help/sexual-reproduction-in-flowering-plants/

Mazziotti, M. (2017). *Impact of Miscanthus x giganteus root exudates on microorganisms involved in the bioremediation of a benzo (a) anthracene-contaminated soil.* University of Lorraine.

Michaels, T., Clark, M., Hoover, E., Irish, L., Smith, A., & Tepe, E. (2022). *8.1 Fruit Morphology.* https://open.lib.umn.edu/horticulture/chapter/8-1-fruit-morphology/

Modification of pea root imprint in response to water shortage (n.d.). INRAE Institutional. Retrieved February 11, 2024, from https://www.inrae.fr/actualites/modification-lempreinte-racinaire-du-pois-reponse-au-manque-deau

Molecular Expressions Cell Biology: Plant Cell Structure-Leaf Tissue Organization. (n.d.). Retrieved February 11, 2024, from https://micro.magnet.fsu.edu/cells/leaftissue/leaftissue.html

Morison, J. I. L., & Lawlor, D. W. (1999). Interactions between increasing CO2 concentration and temperature on plant growth. *Plant, Cell & Environment, 22*(6), 659-682. https://doi.org/10.1046/j.1365-3040.1999.00443.x

Mycorrhiza: Definition, detailed types, diagrams. (s. d.). AquaPortail. Retrieved February 11, 2024, from https://www.aquaportail.com/dictionnaire/definition/9221/mycorhize

Nagwa (n.d.). *Transpiration | Nagwa.* Retrieved February 11, 2024, from https://www.nagwa.com/fr/explainers/202135215651/

Nicotra, A. B., Atkin, O. K., Bonser, S. P., Davidson, A. M., Finnegan, E. J., Mathesius, U., Poot, P., Purugganan, M. D., Richards, C. L., Valladares, F., & Kleunen, M. van. (2010). Plant phenotypic plasticity in a changing climate. *Trends in Plant Science, 15*(12), 684-692. https://doi.org/10.1016/j.tplants.2010.09.008

Nürnberger, T., & Kemmerling, B. (2008). Pathogen-Associated Molecular Patterns (PAMP) and PAMP-Triggered Immunity. In *Annual Plant Reviews Volume 34: Molecular Aspects of Plant Disease Resistance* (pp. 16-47). John Wiley & Sons, Ltd. https://doi.org/10.1002/9781444301441.ch2

Oguchi, R., Hikosaka, K., & Hirose, T. (2003). Does the photosynthetic light-acclimation need change in leaf anatomy? *Plant, Cell & Environment, 26*(4), 505-512. https://doi.org/10.1046/j.1365-3040.2003.00981.x

O'Neill, S. D. (1997). Pollination Regulation of Flower Development. *Annual Review of Plant Physiology and Plant Molecular Biology, 48*(1), 547-574. https://doi.org/10.1146/annurev.arplant.48.1.547

Parenchyma: Definition and explanations. (s. d.). AquaPortail. Retrieved February 12, 2024, from https://www.aquaportail.com/dictionnaire/definition/1272/parenchyme

Parker, J. E. (2000). Signalling in plant disease resistance. In *Molecular Plant Pathology.* CRC Press.

Payette, S., & Filion, L. (2018). *Dendroecology: Principles, methods and applications.* Presses de l'Université Laval.

Plavcová, L., & Jansen, S. (2015). The Role of Xylem Parenchyma in the Storage and Utilization of Nonstructural Carbohydrates. In U. Hacke (Ed.), *Functional and Ecological Xylem Anatomy* (pp. 209-234). Springer International Publishing. https://doi.org/10.1007/978-3-319-15783-2_8

Postel, S., & Kemmerling, B. (2009). Plant systems for recognition of pathogen-associated molecular patterns. *Seminars in Cell & Developmental Biology, 20*(9), 1025-1031. https://doi.org/10.1016/j.semcdb.2009.06.002

Prisa, D. (2023). Role of microorganisms in communication between soil and plants. *Karbala International Journal of Modern Science, 9*(2). https://doi.org/10.33640/2405-609X.3287

Rabinowitch, E. (1949). *Photosynthesis.* US Atomic Energy Commission.

Rathore, J. S., & Ghosh, C. (2018). Pathogen-Associated Molecular Patterns and Their Perception in Plants. In A. Singh & I. K. Singh (Eds.), *Molecular Aspects of Plant-Pathogen Interaction* (pp. 79-113). Springer. https://doi.org/10.1007/978-981-10-7371-7_4

Ratto, F., Simmons, B. I., Spake, R., Zamora-Gutierrez, V., MacDonald, M. A., Merriman, J. C., Tremlett, C. J., Poppy, G. M., Peh, K. S.-H., & Dicks, L. V. (2018). Global importance of vertebrate pollinators for plant reproductive success: A meta-analysis. *Frontiers in Ecology and the Environment, 16*(2), 82-90. https://doi.org/10.1002/fee.1763

Review of plant vs. animal cells (lesson) | Khan Academy. (n. d.). Retrieved February 11, 2024, from https://fr.khanacademy.org/_render

Rhizodeposition. (2023). In *Wikipedia.* https://fr.wikipedia.org/w/index.php?title=Rhizod%C3%A9position&oldid=203911964

Sharma, A., Kohli, S., Khanna, K., M, R., Kumar, V., Bhardwaj, R., Brestic, M., Skalicky, M., Landi, M., & Zheng, B. (2023). Salicylic Acid: A Phenolic Molecule with Multiple Roles in Salt-Stressed Plants. *Journal of Plant Growth Regulation, 42,* 1-25. https://doi.org/10.1007/s00344-022-10902-z

Showalter, A. M. (1993). Structure and function of plant cell wall proteins. *The plant cell*, *5*(1), 9.

SIMON, M. (2009, August 31). *Water, from absorption to transpiration*. Pharmacy course. https://www.cours-pharmacie.com/biologie-vegetale/leau-de-labsorption-a-la-transpiration.html

Sirvydas, A., Kučinskas, V., Kerpauskas, P., Nadzeikiene, J., & Kusta, A. (2010). Solar radiation energy pulsations in a plant leaf. *Journal of Environmental Engineering and Landscape Management*, *18*(3), 188-195. https://doi.org/10.3846/jeelm.2010.22

Smith, W. K., Vogelmann, T. C., DeLucia, E. H., Bell, D. T., & Shepherd, K. A. (1997). Leaf Form and Photosynthesis. *BioScience*, *47*(11), 785-793. https://doi.org/10.2307/1313100

Songer, C. J., & Mintzes, J. J. (1994). Understanding cellular respiration: An analysis of conceptual change in college biology. *Journal of Research in Science Teaching*, *31*(6), 621-637.

sonika (2023, September 11). Fertilization in Plants, Definition, Process and its Types. *Career Power*. https://www.careerpower.in/school/biology/fertilization-in-plants

Thimann, K. V. (1967). Chapter I - Phototropism. In M. Florkin & E. H. Stotz (Eds.), *Comprehensive Biochemistry* (Vol. 27, pp. 1-29). Elsevier. https://doi.org/10.1016/B978-1-4831-9716-6.50009-4

van Lenteren, J., & Ponti, O. M. B. (1991). Plant-leaf morphology, host-plant resistance and biological control. *Symp. Biol. Hung. 39*, 365-386.

Volkov, A. G., Toole, S., & WaMaina, M. (2019). Electrical signal transmission in the plant-wide web. *Bioelectrochemistry*, *129*, 70-78. https://doi.org/10.1016/j.bioelechem.2019.05.003

Wehner, J., Antunes, P. M., Powell, J. R., Mazukatow, J., & Rillig, M. C. (2010). Plant pathogen protection by arbuscular mycorrhizas: A role for fungal diversity? *Pedobiologia, 53*(3), 197-201.

Wilson, C. L., & Just, T. (1939). The morphology of the flower. *The Botanical Review*, *5*(2), 97-131. https://doi.org/10.1007/BF02878180

Yoo, C. Y., Pence, H. E., Hasegawa, P. M., & Mickelbart, M. V. (2009). Regulation of Transpiration to Improve Crop Water Use. *Critical Reviews in Plant Sciences*, *28*(6), 410-431. https://doi.org/10.1080/07352680903173175

Yoshihara, K., & Kumazaki, S. (2000). Primary processes in plant photosynthesis: Photosystem I reaction center. *Journal of Photochemistry and Photobiology C: Photochemistry Reviews*, *1*(1), 22-32.

Zavafer, A., Bates, H., Mancilla, C., & Ralph, P. J. (2023). Phenomics: Conceptualization and importance for plant physiology. *Trends in Plant Science*.

Zhang, H., Zhong, H., Wang, Ji., Sui, X., & Xu, N. (2016). Adaptive changes in chlorophyll content and photosynthetic features to low light in Physocarpus amurensis Maxim and Physocarpus opulifolius "Diabolo". *PeerJ*, *4*, e2125. https://doi.org/10.7717/peerj.2125

Zhang, J. Z., & Reisner, E. (2020). Advancing photosystem II photoelectrochemistry for semi-artificial photosynthesis. *Nature Reviews Chemistry*, *4*(1), Article 1. https://doi.org/10.1038/s41570-019-0149-4

Zhang, Y., Chen, C., Jin, Z., Yang, Z., & Li, Y. (2022). Leaf anatomy, photosynthesis, and chloroplast ultrastructure of *Heptacodium miconioides* seedlings reveal adaptation to light environment. *Environmental and Experimental Botany*, *195*, 104780. https://doi.org/10.1016/j.envexpbot.2022.104780

Zhu, X.-G., Long, S. P., & Ort, D. R. (2010). Improving Photosynthetic Efficiency for Greater Yield. *Annual Review of Plant Biology*, *61*(1), 235-261. https://doi.org/10.1146/annurev-arplant-042809-112206

Printed by Books on Demand GmbH, Norderstedt / Germany